Payments for Farm Environmental Services

by

Paulo Guilherme Salvador Wadt

Grammar Review: Joanna M. Tucker Lima

Text Review: Joanna M. Tucker Lima

Cover: CreateSpace Ltd.

Card catalog

```
W124p  WADT, Paulo Guilherme Salvador
          Payments  for  Farm  Environmental  Services,
       1
          Paulo  Guilherme  Salvador  Wadt,  Plant  City,
       Fl.,  CPS.  2013.
          103p.
          ISBN(10):  1484977092
          ISBN(13):  978-1484977095


          1.  Farm  Management.  2.  Soil  Conservation.
       3.  Environmental  Indicators.  4.  Land  Use.  5.
       Sustainable  Development.  I.  Title.

                                        CCD  631.1+633
```

DEDICATION

In 2010, I met a family of farmers living in Tarauacá County, Acre State, Brazil.

I do not remember anyone's name, but I cannot forget how they lived in poverty on very fertile soils.

This family was living there

Without access to education or health services and without hope.

Living but forgotten by everybody.

Living in a place where everyone said the land was sustainable, but I know no one who wants live there.

I dedicate this book to them.

CONTENTS

ACKNOWLEDGMENTS

My thanks go out to the National Counsel of Technological and Scientific Development (CNPq) and it Science Without Borders Program.

To the Brazilian Agricultural Research Corporation (Embrapa) and the University of Florida (UF) that presented me with new ideas to ponder.

To Dr. Amr Abd-Elrahman and Joanna M. Tucker Lima, who each one in his/her own way helped me along the way I traveled.

To friends and colleagues who withstood the many moments I had to say "no" when they needed my help.

To my parents who have fun far from me, and

Above all, thanks to those who built up walls along the way that I followed, enabling me, after climbing the walls, to see so much farther than I could see without them, and

To the city of Plant City, Florida.

Plant City is a small city, where we can play bingo on holiday evenings.

The Hillsborough Community College houses the University of Florida's Gulf Research Center and is located inside Tampa Bay.

In this place we hear the train pass every day, in the morning, in the afternoon, in the evening and at night, but the trains do not stop here. Sometimes they are going north, sometimes south, sometimes west and sometimes east.

The train is a picture of the past that today is building a new future elsewhere. It does not stop here anymore. It´s like the people who painted a mural in Plant City downtown. They also cannot stop here anymore.

If something on the man's leg looked like a holster, like a holster made to hold a gun like the one that killed so many children in Massachusetts' Newtown School, there would be no problem with the picture. However, if something on the man's leg looked like a sexual organ, it would be better to ban new outdoor paintings in Plant City's Historic Downtown.

If you would like a taste of strawberry cake, here is the best place to eat it. In addition, if you stay here during the Strawberry Festival, you should talk with an old farmer, see a pig run or taste something a vender is showing you even though we can cook it better and more healthfully.

Here we can also walk in the park and watch elementary students practicing soccer, baseball or football any weekday evening.

There are good books in the library, a lot of places to eat and good cars to drive if you are an Enterprise Rent-A-Car customer.

However, if you want a date with someone, you need go to Tampa.

Tampa Nights are only for adults. There are a lot of places to go, but I like one: Ybor City.

If you meet an old mariner in Ybor City, he is probably a liar. He probably wants to sell you something. However, if you want to have fun, then you will have fun. Movies, dinner, drinks and many beautiful people. I like that.

If you want to go to the beach, you don't need to go all the way to the Florida Keys, because there is no place like Clearwater Beach. This is the place to go.

Plant City is special because of all this. If paradise can be found on earth, it is here. On the other hand, if it is not here, then God should use it sometimes if it gets too busy up in heaven.

Because, everyone who lives here is a very special person.

PREFACE

This book aims to expand the approach to environmental service payments. The book discusses two central ideas: (1) environmental services can be linked to agricultural production systems and management practices to combat poverty, and (2) poverty eradication should precede the provision of environmental services.

Neither of these ideas is necessarily new. The connection between environmental services and agricultural production systems is evident in payment programs for agro-environmental services in most major economies. Meanwhile, poverty eradication policies and the fight against poverty also exist in many developing countries even without being linked to environmental services.

Still, this work brings innovative ideas to these two central themes, proposing a range of simple indicators integrated at the farm level, where processes related to assessing the sustainability and quality of land use take place. These indicators follow clear objectives, incur low costs, and are easily applied using either computer systems or more conventional approaches.

This book also deals with sustainability, seeking to recover the application of its most basic concept: how to ensure quality of life for both current populations and future generations.

Much is left to do, and the proposed method needs to be made compatible with geographic database systems offered through the Internet. This should have been chapter six of this book, but the task was beyond my current capacity and available time.

I hope that this book provides a holistic view of the issues that are critical to promoting economic growth in our world of limited natural resources. I believe this can be accomplished using simple indicators based on deterministic processes that merely reflect a small part of the whole system, which in itself is a contradiction and a challenge.

PAULO G S WADT

1 HARVESTING CROPS AND ENVIRONMENTAL SERVICES TOGETHER

"In the history of occidental civilization, the ability to philosophize comes before of the ability to do science" (WEDU 2013)
Ergo, if we want make science without debate, there is something wrong

Ecosystem Services are processes by which the environment produces resources or benefits that we often take for granted, such as clean water, timber, habitat for fisheries, and pollination of native and agricultural plants (Ecological Society of America, 2013).

Different forms of land use can generate a variety of environmental services. For instance, land uses with high levels of tree cover can help regulate water flow in a watershed and reduce the risk of catastrophic flooding or landslides. Governments worldwide have created various types of parks and protected areas managed mainly for ecosystem protection and recreation. These protected areas allow conservation of endemic flora and fauna, biodiversity and ecosystem function. Many protected areas incorporate the exploitation of natural resources, such as "Extractive Reserves" in the Brazilian Amazon (Maciel *et al.*, 2010), or cater to "ecotourism"– responsible travel to natural areas that conserves the environment and sustains local people's well-being– like the National Parks in China (Wang *et al*, 2012).

Farms can provide a variety of environmental services ranging from the regulation of hydrological flows to biodiversity conservation and carbon sequestration (Pagiola *et al.*, 2005). In general, environmental services can be classified as (1) supporting services that include nutrient cycling and primary productivity, (2) provisioning services that include food, fiber, fuel and water, (3) regulating services that include climate regulation and flood control, and (4) cultural services that include recreational, spiritual and aesthetic uses, and each service works at one or more geographic scales (Mann *et al.* 2012).

In Europe, farming landscapes generate a broad range of ecosystem services. These landscapes – principally those with a fine-grained mosaic and low-intensity/extensive production systems – have evolved over centuries alongside humans and provide semi-natural habitat for many endangered and rare species that depend on continued management for their preservation (Brady *et al.*, 2012).

Environmental services extend beyond the cultivable areas within farms. Perhaps the United States (U.S.) has the longest tradition of payments for environmental services on farms, evidenced by its agro-environmental programs (Classen *et al.*, 2008). Prompted by drought, dust storms, and economic depression, the U.S. government began assisting farmers with soil conservation in the 1930s. Since then, they have relied primarily on voluntary payment programs to encourage soil conservation and other improvements to agro-environmental performance.

At present, environmental concerns in the U.S. focus mainly on water quality (surface and ground), wildlife habitat and soil quality (preserving soil productivity), with less attention to air quality, carbon sequestration and energy conservation (Classen *et al.*, 2008). In the United Kingdom, environmental service programs are designed to encourage farmers to manage their lands for improved water quality and reduced soil erosion, improved conditions for farm wildlife, maintenance of soil productivity, enhancement of landscape functions, and protection of historic features (Dobbs & Pretty, 2008).

In all of these programs, payments are made directly to farmers if they meet a series of requirements for environmental programs. Farmers are chosen because they are responsible for how the land is used. If farmers, however, do not receive any compensation for such environmental services they tend to ignore environmental considerations when making land use decisions.

If the conservation practice needed to supply an environmental service is not adopted, the remedial measures needed to fix the environmental damage or loss of benefits are often imperfect and expensive–often far more expensive than preventive measures. Nonetheless, regulatory approaches are extremely difficult to enforce and may impose high costs on poor land users.

In this book, farm describes all tracts of land outside urban areas that are exploited or cultivated for the purpose of agricultural or forest production, devoted to raising and/or breeding domestic animals, or areas of water devoted to raising, breeding, or production aquatic animals. A farm needs to be a continuous area where a single entity is responsible for land use decisions. If farmers make the decision to transform their farm into an environmental service provider, they should receive compensation for environmental services.

Technically, the "farmer" can be an individual, a company or a union, as long as it is someone who makes decisions regarding land use practices and supports the financial costs of these decisions.

In this way, payments for environmental services provide a direct incentive to farmers, encouraging them to adopt land uses that follow conservation practices linked to environmental services, resulting in more socially optimal land uses.

Some authors (Pagiola *et al.*, 2005; Maciel *et al.* 2010) have suggested that environmental service payments can be used to formulate poverty alleviation policies. However, attempts to do this have not produced positive results. For example, the World Bank evaluated the effects of 2.6 billion dollars spent on 289 conservation projects in 75 countries between 2002 and 2011. The investment has served to increase the area of reserves around the world, but in general, failed to prevent the degradation of biodiversity and produced little or no increase in the standard of living for local populations (Campêlo, 2013).

Investments are needed to promote changes in local economies, but are currently insufficient, as globalization does not allow for the same intensity of free movement for goods, technology, labor, and capital between countries or regions. Under current globalization rules, capital can move quickly between countries, but goods and technology have restrictions and labor does not move at all.

Some examples– On the southern border of the U.S., capital moves rapidly between the U.S. and Mexico, even financing the illegal drug and weapon markets: drugs move from Mexico to the U.S., and weapons move from the U.S. to Mexico, but people migrate across the border only with difficulty.

Another example is the Kaesŏng Industrial Region in North Korea. North and South Korea are in war state since 1950. An armistice in 1953 committed both to a cease-fire, but the two countries remain officially at war because a formal peace treaty was never signed. (Wikipedia, 2013).

However, the Kaesŏng Industrial Region in North Korea is a special administrative industrial region of North Korea operated as a collaborative economic development with South Korea.

The park is located six miles north of the Korean Demilitarized Zone. The park allows South Korean companies to employ cheap workers that are educated, skilled, and fluent in Korean, whilst providing North Korea with an important source of foreign currency (Wikipedia, 2013). This occurs with two hostile countries that make goods for the global market with inflow of international capital across their borders.

A recent factory building collapse in Bangladesh that killed more than 1,100 workers (Greenhouse, 2013) is a more recent example of capital moving into poor countries to make goods for the global market, but this move has failed to improve workplace conditions or reduce social inequalities.

Sometimes, even under strict immigration laws, the transfer of labor occurs for short periods to meet immediate interests of big capital, but prevailing labor rights in the destination country offer no guarantees for these workers. For example, large-scale farmers, like those harvesting sweet Vidalia onions in the southern U.S., typically exploit temporary Mexican migrant workers. Working for hours on end under the punishing sun, onion-pickers are said to be crowded into squalid camps, driven to work without breaks and even cheated of wages (Bronner, 2013).

Migrant and seasonal farmworkers are one of the most underserved and understudied occupational populations in the U.S., even though they work in one of the most hazardous occupations in the country (Riley, 2002). Their situation reflects a way of life more common to the nineteenth century than the twenty-first, despite working in the world's largest economy.

Especially in under-developed countries, therefore, the payment of environmental services is likely insufficient to change this scenario in rural areas. Furthermore, anti-poverty policies and strategies can be linked to environmental services only in a subsidiary way, because continuous environmental services cannot be permanent within time and space without supporting sustainable social organizations.

If payments are necessary to convince service providers to adopt required environmental practices, but those same payments are unable to change the lives of those who work in the field, then an alternative is to make farm sustainability a pre-requisite for environmental service providers.

Although "sustainability" has become a remarkably popular word, huge challenges confront those trying to put it into practice. I propose an objective way to measure farm sustainability, which is needed to qualify environmental service providers to receive payments for their services.

The absence of sustainability implies the need for the farmer to make corrections, adjusting their production system before they can operate as an environmental service provider. This means that regional policies on social and economic development should come either before or together with environmental policies and never afterwards.

Another difference in the proposed method is that it focuses on production systems and does not deal solely with non-productive areas. Environmental service payments are designed to suit both demands for productive areas as well as all the other processes related to sustainable land use in non-productive areas within farms.

Also, the method discussed here diverges from agri-environmental services provided by American farmers who bid in a number of ways to apply to the Conservation Reserve Program (CRP). In the U.S., the CRP is a program that resembles environmental service payments (Classen *et al.*, 2008):

Applicants to the CRP program can make a bid to receive annual rental payments for lands that fall below a certain maximum established for the field they are proposing for retirement. This maximum rate or "bid cap" is a function of the county's average cropland rental rates and the productivity of the soils in the field being offered for enrollment. In the CRP program, the Environmental Benefits Index (EBI) is used to rank contracts (offered by farmers) in terms of environmental gains and costs. As the annual rental payment declines, the overall Environmental Benefits Index (EBI) score rises.

Farmers who pay the full cost of establishing ground cover (e.g., grass or trees), rather than accepting cost sharing, also receive additional EBI points. The number of points awarded for the EBI wildlife factor largely depends on the established cover type. If covers that are better for wildlife, such as trees or mixed native grasses, are more expensive to establish, then cover selection could be an integral part of a producer's overall bid.

Finally, producers can affect their EBI score by selecting tracts of land with inherent characteristics that yield higher EBI scores. For example, land with a higher erodibility index will receive more soil erodibility points. This means that farmers can seek to acheive the best combination of factors that will raise their EBI score.

Under these conditions farmers can receive payments even when the farm in its entirety fails to provide environmental services.

In the process proposed in this book, the whole farm is evaluated as a provider of environmental services by integrating all productive and non-productive spaces into a single value for environmental services.

To be compatible with the capability of land use, practices vary according to the farm's technological level to account for large regional differences in access to technology and financial capital.

The proposed method takes into account all these limitations, so as not to confuse agricultural techniques required for appropriate use of the soil with true environmental services, as is possible in the U.S. program.

In addition, the method proposed here meets the requirement for a standardized, transparent assessment of goods and services that can form the basis for decisions regarding the amount of money paid for environmental services (Haaren, *et al.*, 2012). Another characteristic is that the payment is based on a competitive system that involves three main agents: farmer, financial agent and auditor.

The farm agent is responsible for provision, planning and implementing the environment service. The financial agent is the private or public organization responsible for payments related to contracted environment services. Finally, the auditor works to resolve potential conflicts, produces audits and makes information accessible to other agents.

The new system also allows payments to be made through different means. Payments can take the form of lower interest rates for financing, different tax rates on marketable products or differential pricing for products on the regular market. A unique strength of the proposed method is that environmental service payments must be associated with a farm product, such as organic produce, which is offered on the market at a differentiated price, and payments are made to farmers who adopt organic practices in their production systems.

Environmental service payments are designed to encourage farmers to protect and enhance the environment on their farmland. It pays farmers in return for a service–that of carrying out agro-environmental commitments that involve more than the usual every-day good farming practices (European Commission Directorate General for Agriculture and Rural Development, 2005).

Farm Environmental Services (FES) are associated with a land use intensity that is lower than land use capability and with sustainable land use, which promote water, soil and biodiversity conservation for future generations as well as quality of life to farmers and their employees. The outputs generated by FES need to produce both environmental services and social benefits-in other words, harvesting crops and environmental services together. In the next chapters, we discuss:

a) A mathematical formula to quantify the value of farm environmental services (chapter two);

b) A method to assess farm environmental quality (chapter three);

c) A method to define the recommended land use based on capability of land use (chapter four) and;

d) A procedure to evaluate the magnitude of environmental services based in the adequacy of recommended land use (chapter five).

The whole process results in indicators that the common layman can easily interpret and understand. It can also help identify exact details of each process involved and pinpoint where interventions can be made to improve farm performance and sustainability, including the techniques needed to implement the appropriate land use practices.

2 HOW TO CALCULATE THE VALUE FOR ENVIRONMENTAL SERVICES ON FARMS

"Mississippi originally refused to ratify the 13th amendment to United States Constitution in 1865. Despite that, there was a Senate resolution in 1995 that was passed by both the Mississippi House and Senate. That 1995 resolution was never sent to the Office of the Federal Register. That mistake was only corrected on Feb. 7, 2013: 148 years after the rest of the country abolished slavery. After 2013, slavery is no longer legal in Mississippi".
If the abolition of slavery was difficult in such recent times, we need to imagine how difficult it was to do 15 decades ago.

The Environmental Service Payment Value (ESPV) is the monetary value paid to farmers in compensation for his efforts to produce food or other agricultural products under sustainable conditions.

The ESPV is based on three main variables: (1) the Farm Environmental Quality (FEQ) index; (2) the average of the Adequacy of Recommended Land Use (ARL) index for each land unit of the farm; and (3) the difference between the Opportunity Cost of Current Land Use (C) and the Opportunity Cost of Recommended Land Use (R) for each land unit of the farm (Figure 2.1):

If $FEQ \geq 0.7$, then:

$$ESPV = ((3.33 \times (FEQ - 1)) + 1) \times \sum ((ARL_{(i)} / 5) \times A_{(i)} \times (R_{(i)} - C_{(i)}))$$

Else, $ESPV = 0$

Where "i" represents each land use unit inside the farm and "A " is the territorial size of each land use unit, in hectares.

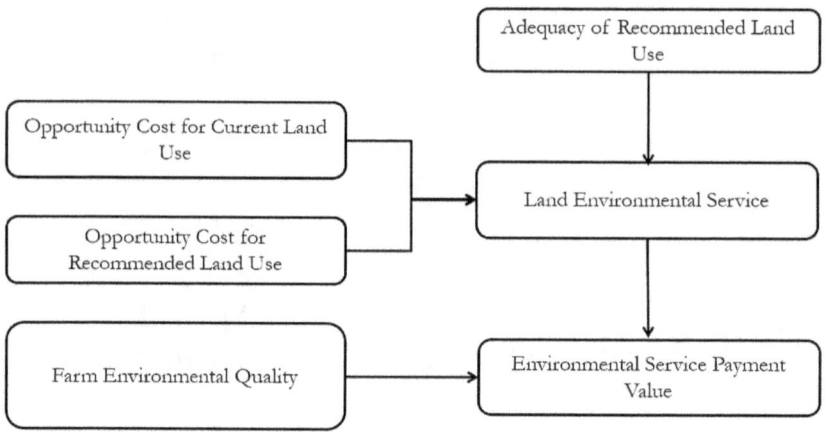

Figure 2.1. Environmental Service Payment Value.

The Farm Environmental Quality (FEQ) index will be discussed in chapter 3 and is a non-dimensional value that corresponds to the farm quality index. The FEQ index indicates whether or not the farm is meeting the minimum conditions for sustainable land management and varies from zero to one, where one is the maximum sustainable value and zero is the minimum sustainable value. If the FEQ index is less than 0.7, the farm is unsustainable and no payments are possible until farm production activities are adjusted toward more sustainable practices. If the farm is sustainable (FEQ $\geq$ 0.7), it can qualify as an environmental service provider.

The Adequacy of Recommended Land Use (ARL) index will be discussed in chapters 4 and 5 and measures the sustainable use of the land. It is a non-dimensional value applied to each land unit and represents the difference between the recommended land use compared with the current land use. Values range from -5 to +5, where ARL greater than 0 means that the current land use is less intense than the recommended land use and ARL less than 0 indicates that the current land use is more intense that the recommended land use.

The expression $(ARL_{(i)} / 5) \times (R_{(i)} - C_{(i)})$ can result in a negative or positive value. It will be negative if the ARL index is negative (i.e., when land use is unsustainable) and positive if the ARL index is positive (i.e., when land use is sustainable).

Only situations where R is greater than or equal to C are valid, although R less than C is possible when less intensive land use implies greater income. If R is less than C, the $(ARL_{(i)} / 5) \times (R_{(i)} - C_{(i)})$ must be equal to zero, because no compensation is due when less intensive land use is more profitable than more intensive land use.

The variable "A" is weighted for each land use unit inside the farm in relation to the entire land area. The farm will only qualify as an environmental service provider if the majority of its territory is under sustainable use, and not just a field or small fraction of the farm.

The opportunity cost quantifies the loss of potential gain from a particular intensive land use when it is replaced by another less intensive land use. Opportunity costs are not restricted to monetary or financial costs: the real costs of missed output, lost time, forgone pleasures or the sacrifice of any other benefit that provides utility should also be considered. Thus, opportunity cost depends not only on local county-wide conditions, such as infrastructure, transportation and land prices, but also on conditions inside the country and between countries.

In this way, the opportunity cost may turn into a powerful tool for public policy, through the governments' ability to drive farmers' adoption of agricultural practices by manipulating values associated with different land uses in their own country.

We can then define the opportunity cost associated with each land use. This value represents the maximum value possible, and depending on the FEQ and the ARL index, it will either be equal to or less than a fixed maximum. The maximum payment possible for the land use is equal to the opportunity cost of the recommended land use minus the opportunity cost of the current land use. A lower FEQ or lower ARL index will decrease the ESPV.

The FEQ index is assessed at the farm scale, but it can also be aggregated at larger scales-Regional Environmental Quality (REQ), such as the bay, county or regional scale (Figure 2.2). At the aggregated scale, FEQ can help unsustainable farms receive payments for environmental services when other farms exhibit high environmental value at the bay, county or regional scale.

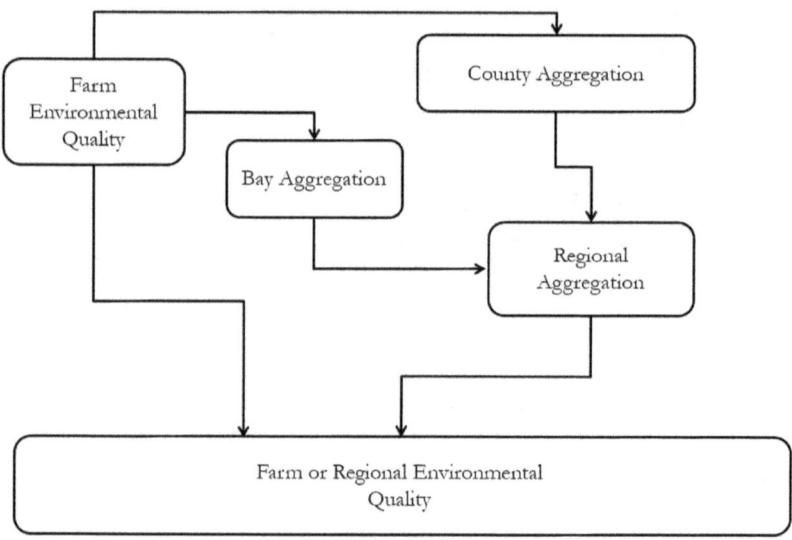

Figure 2.2. Alternative paths to aggregation of farm environmental quality.

If the FEQ is to be aggregated at the bay, county or regional scale, the formula should consider the territorial extent of each farm:

In summary, FEQ and ARL indices need to be determined for each farm and land unit. To do this, an experienced contractor or appropriate financial or governmental agency are needed to evaluate and define the values for R and C.

All the procedures and agencies involved need to work under competitive conditions: financial agencies need to maximize environmental benefits per monetary unit of expenditure, and the farm extension agent needs to offer the maximum environmental benefit to be attractive. Finally, the participation of the extension agent is always voluntary.

The ESPV as presented here is only a suggested value for the payment of farm environmental services. The definitive value will be negotiated between the financial agent and the farmer in consultation with a third party auditor (Figure 2.3).

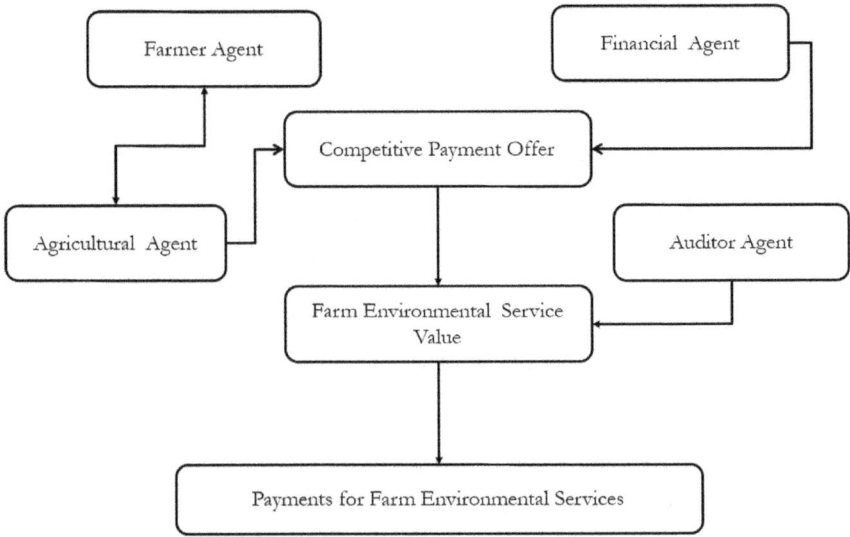

Figure 2.3. Agents involved in Payments for Farm Environmental Services.

Free competition between farmers and financial agents can create the conditions for achieving maximum environmental services with minimal financial resources. In addition, although the FEQ and ARL indices are based on the same indicators to allow comparisons among different farms or regions, the opportunity cost allows differentiated policies to minimize any differences between regions, countries or nations.

3 FARM ENVIRONMENTAL QUALITY

"The Inflationary Universe Theory says that the universe swelled tremendously, going from subatomic size to something as large as the observable universe in a fraction on of a second 13.8 billion years ago" (Hinnant and Borenstein, 2013). If we are able understand this, then we can also understand just how singular life is on this planet.

The Farm Environmental Quality index (FEQ) is a comprehensive measure of farm performance in terms of its environmental quality.

The purpose of the FEQ index is to measure the degree of sustainable development occurring on farms of different sizes, locations and with different production processes. For the purposes of this method, a farm is defined as a tract of land outside urban areas that is cultivated for the purpose of agriculture, forest production or forest extractivism.

The FEQ index is represented by a unique value between zero and one, which represents the overall environmental quality of the entire farm.

This index is calculated based on twenty-six performance indices, where each has the same scale of 0 to 1. These twenty-six performance indices are grouped into five Dimension Categories (DC): Management Dimension, Cultural Dimension, Ambient Dimension, Economic Dimension and Landscape Dimension. Each DC carries the same weight in the FEQ calculation.

The number of performance indices in each DC varies, but each DC requires a minimum number of performance indices to be included in the FEQ index, which should represent more than 75% of the performance indices of each DC (Figure 3.1).

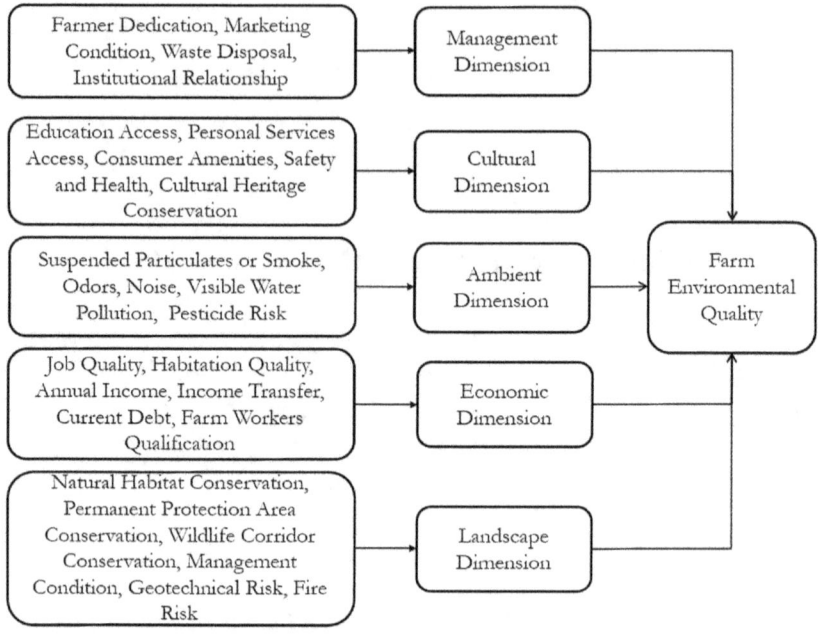

Figure 3.1. Relationships between the performance indices.

Each performance index, DC index or FEQ index are the only mathematical resources readily available to transform indicators derived from different units and different physical dimension into a self-comparable value and express them as a dimensionless quantity (quantity without an associated physical unit), in a scales of 0 to 1. These indices are also called Utility Coefficients (UC) (Andreoli and Tellarini, 2000).

A UC value lower than 0.7 denotes to an unsustainable process and a value higher than 0.7 suggests a sustainable process. As such, one can assess the degree of sustainable development from the farm's UC value (FEQ), for each of the dimensions (DC) or for a specific process (performance index), irrespective of the understanding of biological variables or physical quantities involved. This facilitates the comparison of the performance of different processes and makes communication easier, since knowing whether the value is greater than or less than 0.7 is sufficient to determine the degree of sustainability of each process .

The FEQ method was adapted from the "Environmental Impact Assessment System of New Rural Activities" (Rodrigues e Campanhola, 2003). Changes to the original method include:

a) Change in the time trends index within the performance indices.

The original method measures environmental impacts using time trends. Time trends are consistent with interpreting sustainability as the ability to continue. Sustainability assessments are made in terms of the direction and degree of measurable changes in each system property. The system is considered sustainable if there is no negative trend among selected system properties (Smith and McDonald, 1988).

However, time trends incorrectly measure sustainability between farms with different resource demands, market structures and technology: farms with more recent deforestation may appear to have more sustainable soil management, because slashing and burning of the forest still positively influences soil fertility, while older farms can appear less sustainable, despite the use of more suitable land management practices (Avila, 2006).

In the new approach, all indicators were altered to measure the performance of related processes. Hence, each indicator was assigned a performance curve, which split each process into lower performance (unsustainable) and higher performance (sustainable).

b) Removal of indices inappropriate to the farm scale.

The original method uses various indices that include soil fertility indicators, like soil pH, soil ex-changeable cations, and soil exchangeable aluminum. These indicators vary widely among different land use units within the farm, and none is able to represent the entire farm area. Some processes related to these indicators were eliminated from the new method but have been incorporated into a complementary method that evaluates the recommended land use (Delarmelinda, 2011) and is described in the chapter four.

c) Removal of indices with high dependency on processes occurring outside the farm.

Indices that measure air quality, like concentrations of carbon oxides, sulfur oxides or oxides of nitrogen in the farm's air were removed, because in all situations considered for small farms, the processes occurred outside of the farm boundaries.

In addition, indices such as species extinction risk and incidence of endemic disease vectors focus on processes that take place outside the farm, and most were only appropriate for very large farms or landscapes at the regional scale (Avila, 2006).

d) Change of various local situation indices to a weighted average condition index that takes into account the territorial distribution of lands within the farm.

In the original method, indicators like fire risk and geotechnical risk were computed for the entire farm with no accounting for variation across within-farm land units. This was corrected by computing these indicators for each land unit and then adding them together using weighted averages to calculate processes at the whole farm level.

While the original method uses sixty-two indicators to measure environmental impact (Rodrigues and Campanhola, 2003), the new approach uses only twenty-six indicators, although retaining the same number of DCs. In the new approach some processes are spatially dependent, whereas another processes were maintained without spatial dependency.

The FEQ index presented here is very different from the original method described by Rodrigues and Campanhola (2003), although all remaining indicators measure the same processes as the original method. Both methods also use UC and consider 0.7 to be the cut-off value between sustainable and unsustainable processes.

Summarizing:

FEQ = (MDim + CDim + EDim + ADim + LDim)/5

Where:

FEQ is a dimensionless index of Farm Environmental Quality and MDim, CDim, EDim, ADim and LDim are dimensionless indices of Management Dimension, Cultural Dimension, Ambient Dimension, Economic Dimension and Landscape Dimension, respectively.

In addition:

$MDim = (Mfd + Mmc + Mwd + Mir)/m$

Where:

Mfd, Mmc, Mwd, Mir are dimensionless performance indices of Farmer Dedication, Marketing Condition, Waste Disposal and Institutional Relationship, respectively; and

m = number of performance indices used to compute MDim, where "m" is always 3 or greater.

$CDim = (Cea + Cps + Cca + Csh + Cch)/c$

Where:

Cea, Cps, Cca, Csc, Cch are dimensionless performance indices for Education Access, Personal Services Access, Consumer Amenities, Safety and Health; and Cultural Heritage Conservation, respectively; and

c = number of performance indices used to calculate CDim, where "c" is always 4 or greater.

$EDim = (Ejw + Ehq + Eai + Eit + Ecd + Efw)/e$

Eqw, Eiq, Eai, Eit. Edc e Efw are dimensionless performance indices of Job Quality, Habitation Quality, Annual Income, Income Transfer, Current Debt and Farm Workers Qualification, respectively; and

e = number of performance indices used to calculate EDim, where "e" is always 5 or greater.

$ADim = (Asp + Aod + Ano + Avp + Apr)/a$

Where:

Asp, Aod, Ano, Avp, Apr are dimensionless performance indices of Suspended Particulates or Smoke, Odors, Noise, Visible Water Pollution and Pesticide Risk, respectively; and

a = number of performance indices used to calculate ADim, where "a" is always 4 or greater.

$$LDim = (Lnh + Lpp + Lwc + Lmc + Lgr + Lfr)/l$$

Where:

Lnh, Lpp, Lwc, Lmc, Lgr and Lfr are dimensionless performance indices of Natural Habitat Conservation, Permanent Protection Area Conservation, Wildlife Corridor Conservation, Management Condition, Geotechnical Risk and Fire Risk, respectively; and

l = number of performance indices used to calculate LDim, where "l" is always 5 or greater.

The calculation of performance indices inside each Dimensional Category involves indicators with different physical scales or physical properties (Table 3.1). The calculations involved in each performance index are described below.

Table 3.1. Performance indices and indicator indices in each Dimension Category of the Farm Environmental Quality (FEQ) index.

Performance index	Indicators	Indicator quantity	Code
Management Dimension (MDim)			
Farmer Dedication	local residence, dedication to farm business, farm business training, family engagement, formal notability and formal planning system	6	Mfd
Marketing Condition	Direct sales, agro-processing at rural or union level, post-harvest handling and storage, own transportation; own advertising and marketing, own brand, production chain and sales to other producers.	8	Mmc
Waste Disposal	Waste sorting, composting, sanitary disposal, recycling and final treatment.	5	Mwd
Institutional Relationship	Regular technical assistance, cooperative association/union participation, technological association, advanced level technical assistance, manager training and farm workers training	6	Mir
Cultural Dimension (CDim)			
Education Access	Educational degree of the farm workers (no elementary school, elementary school, high school, college professional degree, upper professional degree).	1	Cea
Personal Services Access	Access to basic services by farmer's family (potable water, electricity, telephone, transportation and health services near work place)	1	Cps
Consumer Amenities	Access to consumer amenities by farmer's family (gas or electrical stove, refrigerator, television, internet access, own vehicle, washing machine and microwave oven).	1	Cca
Safety and Health	Farm workers exposure to occupational safety risks like explosion hazards, flammable products, electrical or ionizing products and farm workers exposure to occupational health hazards such as loud noises, physical vibration, intense heat or cold conditions, high humidity, chemical risks and biological risks	2	Csh

Continues…

Continuation of the table 3.1

Cultural Heritage Conservation	Conservation status (excellent, good, fair or poor) of cultural heritage (historical, artistic, archaeological or speleological)	1	Cch

Economic Dimension (EDim)

Job Quality	Absence of legal employment contracts; absence of access to basic insurance program; absence of basic rights at work and liquid wage below national minimum wage.	4	Ejq
Habitation Quality	Building standards used in construction of the farmer's house and average number of people per bedroom	2	Ehq
Annual Income	Tendency to increase, maintain or decrease in security, income stability and annual income	3	Eai
Income Transfer	Relationship between annual net income and annual wages paid	1	Eit
Current Debt	Sum of all financial obligations that need to be paid over the next year in relation to total annual net income during the same period	1	Ecd
Farm Workers Qualification	Demand for manual labor jobs, skilled laborers, middle level and higher level workers living on the property, in the town or in another city	1	Efw

Ambient Dimension (ADim)

Suspended Particulates or Smoke	The first four processes are associated with occurrence of the phenomenon on exact place, local or regional scales, under	1	Asp
Odors	different intensities varying from weak to	1	Aod
Noise	uncomfortable or unbearable, and over a 1	1	Ano
Visible Water Pollution	to 12 month period.	1	Avp
Pesticide Risk	Tendency to increase, maintain or decrease in frequency, variety and toxicity of pesticide use at farm	1	Apr

Continues…

Continuation of the table 3.1

Landscape Dimension (LDim)			
Natural Habitat Conservation	Distribution of and management inside natural habit conservation areas	1	Lnh
Permanent protection areas	Management applied to conservation of permanent preservation areas	1	Lpp
Wildlife Corridor Conservation	Management applied to and fragmentation type of wildlife corridor areas	1	Lwc
Management Condition	Use intensity and management quality inside production unit areas	1	Lmc
Geotechnical Risk	Geological vulnerabilities associated with land use units	1	Lgr
Fire Risk	Fire management practices associated with land use units	1	Lfr

Management Dimension

The Management Dimension (MDim) measures the farmer's administrative and management conditions, which are considered sustainable if management includes the adoption of modern administration techniques, waste management, and a solid relationship with the market and technical assistance. This means that decisions guided by practical experience rather than administrative theory and absence of market or agricultural technology resources will be unsustainable.

The Management Dimension (MDim) includes the average of at least three of four processes: "Farmer Dedication", "Marketing Conditions", "Waste Disposal" and "Institutional Relationship".

The four performance indices (Mfd, Mmc, Mwd, Mir) for MDim use the same logistic model to transform binomial data into a Utility Coefficient (UC) (Figure 3.2), but the curvature of the exponential model varies by 0.375 to 0.750.

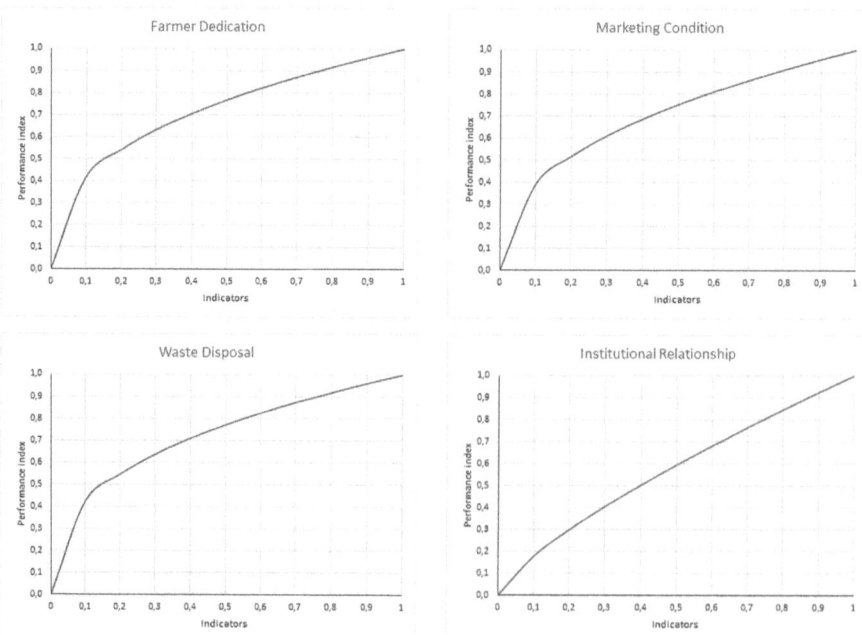

Figure 3.2. Logistic model to compute performance indices for the Management Dimension

Farmer Dedication

This performance index measures the degree of the farmer's commitment to managing his farm. A higher degree of commitment denotes greater business sustainability. Training directed at the main farm business and use of a formal planning system are the two main processes considered in this performance index.

The Farmer Dedication performance index (Mfd) is calculated by the expression:

$$Mfp = ((fdLR + fdED + (3 \times fdCD) + fdFE + fdFN + (3 \times fdFP)) / 10)^{0.38}$$

Where indicators are local residence / farmer living onsite (fdLR), farmer activity exclusively dedicated to farm business (fdED), coaching directed at the main farm business (fdCD), family engagement in the farm business (fdFE), existence of formal notability (fdFN) and use of a formal planning system (fdFP). All indicators are represented by binary values, where 1 denotes occurrence and 0 denotes absence.

Marketing Condition

This performance index measures the marketing conditions related to farm products. The higher the farmer's control over the chain of production processes outside the farm, the greater the business sustainability.

Direct sales to consumer, personal trademark and vertical integration of decisions regarding production process are the three main processes considered in this performance index.

Marketing Condition performance (Mmc) is calculated by the expression:

$$Mmc = ((((3 \times mcDS) + mcOP + mcOS + mcOT + mcOM + (3 \times mcTM) + (2 \times mcVP) + mcSF))/10)^{0.410}$$

Where indicators are direct sales to consumer (mcDS), own processing (mcOP), own storage (mcOS), own transportation (mcOT), own marketing (mcOM), trademark (mcTM), vertical integration of decisions about the production process, with participation in decisions relating to production, storage, processing and sales (mcVP), and sales to other farms (mcSF). All indicators are binary values, where 1 denotes occurrence and 0 denotes absence.

Each process can be directed by the farmer or by a cooperative association/union of farmers.

Waste Disposal

This performance index measures waste disposal management inside the farm. The adoption of adequate waste disposal practices implies a higher degree of farm waste management sustainability.

Sanitary disposal and waste recycling are the two main processes examined in this performance index.

Waste Disposal performance (Mwd) is calculated by the expression:

$$Mwd = ((wdWT + wdOC + (3 \times wdSD) + (2 \times wdRY) + wdET)/10)^{0.375}$$

Where indicators are selective waste treatment (wdWT), organic fertilizer composting (wdOC), sanitary disposal (wdSD), waste recycling (wdRY) and satisfactory disposal of waste to prevent ambient contamination by using end treatment process for farm waste (wdET). All indicators are binary values, where 1 denotes occurrence and 0 denotes absence.

Institutional Relationship

This performance index measures the farm business' stage of technological development. Access to technical information and training implies a higher probability that the farmer's decisions will lead to continuity of the business over the long term.

Access to regular technical assistance and farmer participation in cooperative associations/unions are the two main process considered in this performance index.

Institutional Relationship (Mir) is calculated by the expression:

$$Mir = (((3 \times irRA) + (2 \times irUP) + irTA + irHT (2 \times irMT) + irET) / 10)^{0.75}$$

Where indicators are access to regular technical assistance (irRA), farmer participation in cooperative association/union (irUP), participation in technological association (irTA), technical assistance with high degree of expertise (irHT), manager training (irMT) and employee training (irET). All indicators are binary values, where 1 denotes occurrence and 0 denotes absence.

If the farm has performance indices of 0.7 or greater, then the farm's management will be competitive under different farm business scenarios

.

Cultural Dimension

The Cultural Dimension (CDim) takes into account life quality indicators. Farmer and farm worker access to better quality of life conditions provides a standard of living favorable to the maintenance of production systems on the farm. Higher quality of life for farmers and farm workers means higher sustainability for the farm business.

The Cultural Dimension represents the average of at least four of the five performance indices: Education Access, Personal Services Access, Consumer Amenities, Safety and Health; and Cultural Heritage Conservation.

The five performance indices (Cea, Cps, Cca, Csh and Cch) included in CDim use the same logistic model to transform data into the Utility Coefficient (UC) (Figure 3.3), but the curvature of the exponential model varies by 0.250 to 1.450. These indicators have different data types.

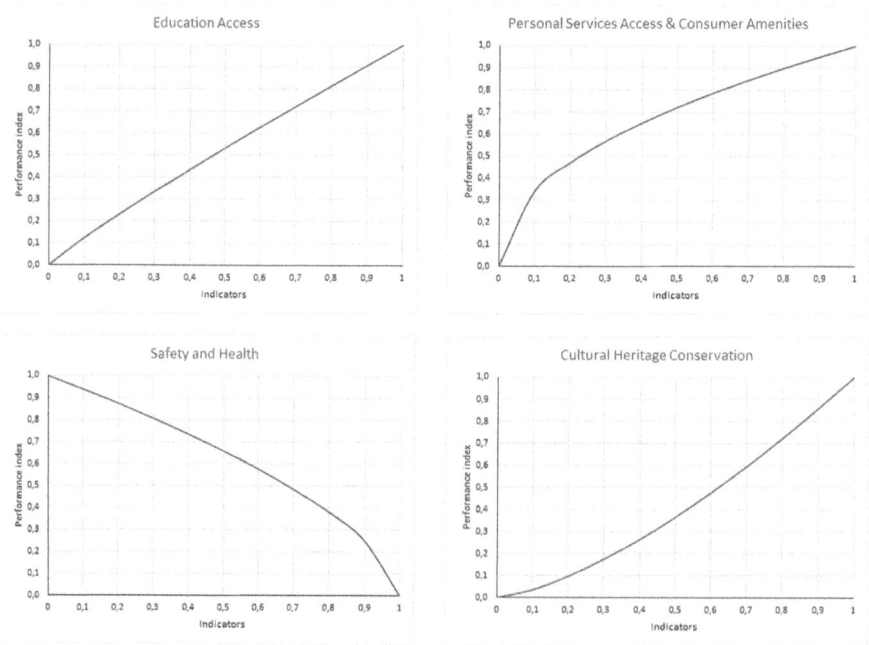

Figure 3.3. Logistic model to compute performance indices for the Cultural Dimension.

Education Access

This performance index measures access to quality education for farm workers. The higher the degree of worker education, the more qualified workers are for the job and the farm business is more sustainable.

Farm workers with an advanced specialization professional degree represent a better condition in this performance index.

Education Access (Cea) is calculated by the expression:

$$Cea = (((eaES + (3 \times eaHE) + (6 \times eaCD) + (10 \times eaPD)) / (10 \times (eaES + eaWS + eaHE + eaCD + eaPD)))^{0.91}$$

Where indicators are the quantity of farm workers with no elementary school (eaWS), elementary school education (eaES), high school education (diploma) (eaHE), a college professional degree (eaCD) and an advanced specialization professional degree (eaPD). All indicators are represented by counts.

Personal Services Access

This performance index measures the farmer family's access to basic services and reflects the farmer family's quality of life.

All processes are given the same weight when computing the service access index.

Personal Services Access (Cps) is calculated using the expression:

$$Cps = (psWP + psEL + psPH + psTN + psHN) / 5)^{0.47}$$

Where indicators include access to potable water (psWP), electricity (psEL), telephone (psPH), transportation services near the work place (psTN) and health services near the work place (psHN). All indicators are binary values, where 1 denotes occurrence and 0 denotes absence.

Consumer Amenities

This performance index measures the consumer amenities acquired by the farmer's family and reflects quality of life.

Access to gas or electrical stove, washing machine and microwave oven are the three main processes included in this performance index.

Consumer Amenities index (Cca) is calculated by the expression:

$$Cca = (caES + (2 \times caRF) + caTV + caIT + caOV + (2 \times caCM) + (2 \times caMO)) / 10)^{0,47}$$

Where indicators are access to consumer amenities by the farmer's family: gas or electrical stove (caES), refrigerator (caRF), television (caTV), internet access (caIT), own vehicle (caOV), washing machine (caCM) and microwave oven (caMO). All indicators are reported as binary values, where 1 denotes occurrence and 0 denotes absence.

Safety and Health

This performance index measures safety and health of farm workers and reflects the risks that workers are exposed to during the regular workday. The higher the number farm workers exposed to inappropriate job conditions, the lower the farm's sustainability.

All of the processes are assigned the same weight when computing the safety and health performance index.

The Safety and Health index (Csh) is calculated by the expression:

$$Csh = (1 - ((shEH + shFP + shEI + shHL + shPV + shHH + shHC + shQR + shBR) / (9 \times FW)))^{0.60}$$

Where indicators are the number of farm workers (FW) exposed to inappropriate occupational safety hazards: explosion hazards (shEH), flammable products (shFP), electrical or ionizing products (shEI); and exposures to inappropriate occupational health hazards: loud noise (shHL), physical vibration (shPV), extreme heat or cold (shHH), high humidity (shHC), chemical risks (shQR) and biological risks (shBR). All indicators are represented by counts.

Cultural Heritage

Few farms have cultural heritage (historical, artistic, archaeological or speleological), but when present, it is necessary to assess their degree of conservation. This performance index is computed only when farms have cultural heritage. Cultural Heritage is classified as being in excellent, good, bad or poor conservation status. Poor conservation is when there is not attention to conservation; excellent, good and badly is when the care to conservation is able to maintain your properties of cultural heritage, good is when there is some loss of your properties and badly is when there is much loss of your properties,

This performance index measures the conservation state of cultural heritage, and reflects the care taken by farmers to conserve their farm's heritage.

All processes have the same weight when computing the cultural heritage performance index.

Cultural Heritage (Cch) is calculated by the expression:

If (chEC + chGC + chBC + chPC) > 0, then:

Cch = ((chEC + (0.7 x chGC) + (0.35 x BC)) / (chEC + chGC + chBC + chPC))$^{1.45}$

Where indicators are the number of cultural heritage items that are in excellent (chEC), good (chGC), badly (chBC) or poor conservation status (chPC). All indicators are represented by counts.

Economic Dimension

The Economic Dimension (EDim) measures the performance of certain economic indicators without a comprehensive economic analysis. These indicators do not measure the financial health or the economic stage of the farm business, but instead employ easy and direct indicators to compare farms with each other.

This dimension is described by the arithmetic mean of at least five of six performance indices: Job Quality, Habitation Quality, Annual Income, Income Transfer, Current Debt and Farm Worker Qualification.

The six performance indices (Ejq, Ehq, Eai, Eit, Ecd and Efw) for EDim use different logistic models to transform data into a Utility Coefficient (UC) (Figure 3.4). These indicators also have different data types.

Job Quality

This performance index measures the quality of relationship between farm workers and the farmer who manages their work contracts. The Job Quality index (Ejq) will be higher when farm workers receive all of the social benefits through their work contracts.

The index has four indicators based on the number of farm employees:

a) With illegal employment contracts, such that the work relationship lacks any written statement of the main terms and conditions of employment (jqWS);

b) Without access to a basic insurance program, such as access to retirement, life or disability insurance programs offered by the government or by the employer (jqIP);

c) Without basic rights at work, like sick pay when away from work due to illness, maternity leave, paternity leave or adoption leave, or parental leave, and a certain number of days paid holiday per year (jqBR); and

d) With wages less than twice the national minimum wage (jqNW).

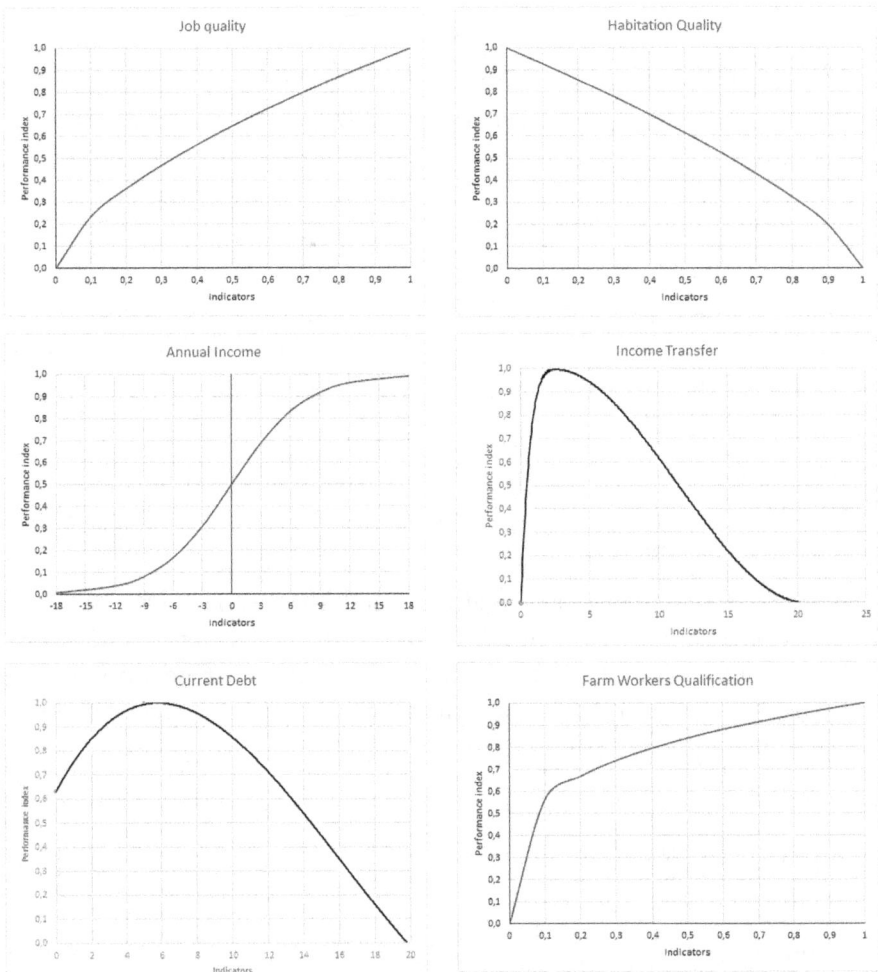

Figure 3.4. Logistic model to compute performance indices for the Economic Dimension.

In the calculation of this performance index one needs to know the total number of farm workers (FW). Each job that lasts at least one month out of the year is included in the calculations, but workers employed for less than one month are excluded.

All processes are assigned the same weight to compute the job quality performance index.

Job Quality (Ejq) is calculated by the expression:

If FW > 0, then:

$$Ejq = ((jqWS + jqIP + jqBR + jqNW) / (4 \times jqNW))^{0.63}$$

Where indicators described above are the number of farmworkers involved in each category.

Habitation Quality

This performance index considers the construction quality of the farmer´s house, when it is located within the farm boundaries; otherwise, if the farmer's home is located outside the farm, this index cannot be assessed.

Habitation Quality (Ehq) will be higher if the farmer's house is comfortable and has an adequate distribution of bedrooms per person living in the house.

The type of construction pattern used for the farmer's house is rated as excellent, good, regular or bad based on whether the home has insulation from heat and cold, water isolation, protection against biological agents and the average number of the people per bedroom.

The Habitation Quality index (Ehq) is calculated by the expression:

If hqAP < hqPH x 3,

Then: Ehq = 0

Otherwise: Ehq = $[1 - [((2 \times hqRoom3) + (3 \times hqRoom4) + (4 \times hqRoom5)) / (8 \times (hqRoom2 + hqRoom3 + hqRoom4 + hqRoom5))] + hqCQ]^{0.7}$

Where hqAP is the sum of bedroom area in the farmer's house, hqPH is number of persons living in the farm house, and the number 3 represents the minimum area of each bedroom (3 m² per person). And, hqRoom2, hqRoom3, hqRoom4, hqRoom5 are the number of persons living in the bedroom with 1 or 2 persons, 3 persons, 4 persons and 5 or more persons, respectively. Finally, hqCQ is the quality of construction used to build of the farmer's house (insulation from heat and cold, protection against biological agents), ranked as follows: 0.5 is excellent, 0.35 is good, 0.17 is regular and 0.0 is badly).

A status of excellent indicates thermal comfort and biological and water isolation; if one of these categories is not present, status is good; if two categories are missing, status is regular and when none of the categories are present, status is labeled as bad.

Annual Income

This performance index considers the evolution of the farmer's annual income generated from farm activities. The index measures the tendency to increase, maintain or decrease security (facility to sell farm products and presence of satisfactory markup - the difference between the cost of a good or service and its selling price), stability (if product prices are competitive and have no other product substitute entering the market) and annual income (farm annual net income).

If all processes tend to maintain the same intensity as the recent past, the performance index will be 0.5. If the tendency is to decrease, the performance index falls below 0.5, and if the quality of the process tends to increase, the performance index will be above 0.5. The performance index is only sustainable when none of the processes tend to decrease and if at least one process shows a tendency to increase.

The Annual Income index (Eai) is calculated by the expression:

Eaivar = [6 x [(aiSECin – aiSECdec) + (aiSTBin – aiSTBdec) + (aiANNin – aiANNdec)]]

$$Eai = 1 / (1 + 0.41^{(0.3 \text{ x Eaivar})})$$

Where aiSECin and aiSECdec represent the tendency for security to increase or decrease, respectively; aiSTBin and aiSTBdec are the tendency for stability to increase or decrease, respectively; and aiANNin and aiANNdec are the tendency for annual income to increase or decrease, respectively. Values for these indices will be one or zero, where one indicates the process is true and zero indicates false. If the tendency is to maintain process intensity, the value will be zero for both increasing and decreasing possibilities.

Income Transfer

This performance index measures the partition of the farm's net income between farm workers and the farmer. The index will be at its maximum when there is equilibrium in the distribution of income between the parts involved in this process. A more sustainable condition is achieved when the farmer's income is twice that of the farm workers'.

Income Transfer (Eit) is calculated by the expression:

If itFI/itWI < 2

Then: Eit = $0.113(\text{itFI/itWI})^3$ - 0.6622 x $(\text{itFI/itWI})^2$ + $1{,}3687$ x (itFI/itWI)

Else: Eit = $0.0003(\text{itFI/itWI})^3$ - 0.0119 x $(\text{itFI/itWI})^2$ + 0.0528 x (itFI/itWI) + 0.9334

Where itFI and itWI are the annual income received by the farmer and the farm workers, respectively. All values are in monetary units.

Current Debt

This performance index measures two different situations: if the farmer has access to financial aid for his/her business and if the farmer's debt is within his/her ability to pay. The ideal situation is when the farmer has annual debts equivalent to 20% of their annual income.

Ecd = 0.0003 x $(\text{cdAI/cdWI})^3$ - 0.0145 x $(\text{cdAI/cdWI})^2$ + 0.1376 x (cdAI/dcWI) + 0.632

Where cdAl and cdWI represent total annual net income and the sum of all financial obligations that the farmer needs to pay over the next year. All values are in monetary units.

Farm Workers Qualification

This performance index measures the quality of the jobs demanded by the farm. The greater the number of farm workers with a higher education degree, the higher the performance index.

Farms that only demand workers with low educational qualifications will not be sustainable.

Farm Workers Qualification (Efw) is calculated by the expression:

$$Efw = (\ (5 \times fwFL) + (3 \times fwCL) + fwOL + (10 \times fwFE) + (5 \times fwCE) + (3 \times fwOE) + (20 \times fwFH) + (10 \times fwCH) + (5 \times fwOH) + (40 \times fwFC) + (20 \times fwCC) + (10 \times fwOC))\ /(40 \times (fwFL + fwCL + fwOL + fwFE + fwCE + fwOE + fwFH + fwCH + fwOH + fwFC + fwCC + fwOC))^{0.25}$$

Where:

$fwFL$ = number of farm workers who live inside the farm and have no elementary education;

$fwCL$ = number of farm workers who live inside the county and have no elementary education;

$fwOL$ = number of farm workers who live outside the county and have no elementary education;

$fwFE$ = number of farm workers who live inside the farm and only have an elementary education;

$fwCE$ = number of farm workers who live inside county and only have an elementary education;

$fwOE$ = number of farm workers who live outside the county and only have an elementary education;

$fwFH$ = number of farm workers who live inside the farm and have a high school diploma;

$fwCH$ = number of farm workers who live inside the county and have a high school diploma;

$fwOH$ = number of farm workers who live outside the county and have a high school diploma;

$fwFC$ = number of farm workers who live inside the farm and have a college education;

$fwCC$ = number of farm workers who live inside the county and have a college education;

$fwOC$ = number of farm workers who live outside the county and have a college education.

In the categories above, farm workers who live "inside the county" refers to workers who live outside the farm.

All values are reported as counts.

Ambient Dimension

The Ambient Dimension (ADim) measures ambient qualities, like air, water and noise on the farm and the surrounding area.

Four of the performance indices use the same logistic model: suspended particulates or smoke, odors, noise and visible water pollution. The fifth performance index that helps define this dimension uses a different logistic model (Figure 3.5). The Ambient Dimension represents the arithmetic mean of at least four of five processes: Suspended particulates or smoke, Odors, Noise, Visible water pollution and Pesticide Risk.

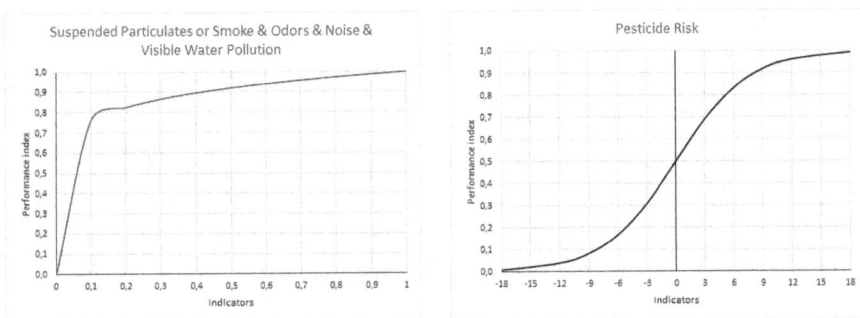

Figure 3.5. Logistic model to compute performance indices for the Ambient Dimension.

For the following four performance indices, site scope indicates that the phenomenon is restricted to the location where it was generated. Local scope means that the phenomenon affects most of the farm. Neighborhood scope specifies that the phenomenon reaches the farm and its surroundings. Finally, regional scope indicates that the phenomenon affects the regional scale, such as the hydrographic basin or the county.

Low intensity means that the phenomenon is present but is not easily perceived; uncomfortable intensity is when the phenomenon is perceived but demands no special measures to relieve discomfort; and unbearable intensity is when the phenomenon reaches a level of discomfort.

All values are reported as counts. If the phenomenon occurs at low intensity for two months of the year and at uncomfortable intensity for one month a year, the number of months per year for each occurrence must be registered in different categories, and other months are recorded as zeros, up to a maximum of twelve months a year.

Suspended Particulates or Smoke

This performance index measures the air quality in terms of air pollutants, such as smoke or suspended particulates. For this model, the division between sustainable and unsustainable depends on the presence of smoke or suspended particulates in one month of the year at low intensity.

The Suspended Particulates or Smoke (Asp) performance index is calculated using the expression:

$$Asp = (1 - (((0.1 \times 0.03 \times spSL) + (1 \times 0.03 \times spLL) + (5 \times 0.03 \times spNL) + (10 \times 0.03 \times spRL) + (0.1 \times 0.05 \times spSC) + (1 \times 0.05 \times spLC) + (5 \times 0.05 \times spNC) + (10 \times 0.05 \times spRC) + (0.1 \times 0.1 \times spSB) + (1 \times 0.1 \times spLB) + (5 \times 0.1 \times spNB) + (10 \times 0.1 \times spRB)) / 12)^{12}$$

Where:

spSL = site scope with low intensity;

spLL = local scope with low intensity;

spNL = neighborhood scope with low intensity;

spRL = regional scope with low intensity;

spSC = site scope with uncomfortable intensity;

spLC = local scope with uncomfortable intensity;

spNC = neighborhood scope with uncomfortable intensity;

spRC = regional scope with uncomfortable intensity;

spSB = site scope with unbearable intensity;

spLB = local scope with unbearable intensity;

spNB = neighborhood scope with unbearable intensity;

spRB = regional scope with unbearable intensity.

Odors

This performance index measures air quality in terms of the presence of unpleasant smells or odors derived from some source of air pollution. For this model, the division between sustainable and unsustainable was marked by the presence of odors in at least one month of the year at low intensity.

The Odor (Aod) performance index is calculated by the expression:

Aod = (1 - (((0.1 x 0.03 x odSL) + (1 x 0.03 x odLL) + (5 x0.03 x odNL) + (10 x 0.03 x odRL) + (0.1 x 0.05 x odSC) + (1 x 0.05 x odLC) + (5 x 0.05 x odNC) + (10 x 0.05 x odRC) + (0.1 x 0.1 x odSB) + (1 x 0.1 x odLB) + (5 x 0.1x odNB) + (10 x 0.1 x odRB)) /12) [12]

Where:

odSL = site scope with low intensity;

odLL = local scope with low intensity;

odNL = neighborhood scope with low intensity;

odRL = regional scope with low intensity;

odSC = site scope with uncomfortable intensity;

odLC = local scope with uncomfortable intensity;

odNC = neighborhood scope with uncomfortable intensity;

odRC = regional scope with uncomfortable intensity;

odSB = site scope with unbearable intensity;

odLB = local scope with unbearable intensity;

odNB = neighborhood scope with unbearable intensity;

odRB = regional scope with unbearable intensity.

Noise

This performance index measures noise pollution and assesses the presence of noise or sounds not strictly related to natural phenomena, such as vehicles, machinery operations and other sounds resulting from human activities. For this model, the limit between sustainable and unsustainable was based on the presence of noise during at least one month of the year at low intensity.

The Noise (Ano) performance index is calculated using the expression:

Ano = (1 - (((0.1 x 0.03 x noSL) + (1 x 0.03 x noLL) + (5 x0.03 x noNL) + (10 x 0.03 x noRL) + (0.1 x 0.05 x noSC) + (1 x 0.05 x noLC) + (5 x 0.05 x noNC) + (10 x 0.05 x noRC) + (0.1 x 0.1 x noSB) + (1 x 0.1 x noLB) + (5 x 0.1x noNB) + (10 x 0.1 x noRB)) /12) [12]

Where:

noSL = site scope with low intensity;

noLL = local scope with low intensity;

noNL = neighborhood scope with low intensity;

noRL = regional scope with low intensity;

noSC = site scope with uncomfortable intensity;

noLC = local scope with uncomfortable intensity;

noNC = neighborhood scope with uncomfortable intensity;

noRC = regional scope with uncomfortable intensity;

noSB = site scope with unbearable intensity;

noLB = local scope with unbearable intensity;

noNB = neighborhood scope with unbearable intensity;

noRB = regional scope with unbearable intensity.

Visible Water Pollution

This performance index measures water quality and specifically considers the presence of water pollutants visible to the naked eye or detected by other techniques. Only natural phenomena that affect water quality can be discarded. Processes in the water that are invisible but cause environmental problems, such as fish kills, are also considered. For this model, the division between sustainable and unsustainable was based on the occurrence of water pollution during at least one month of the year at low intensity.

The Visible Water Pollution (Avp) performance index is calculated by the expression:

Avp = (1 - (((0.1 x 0.03 x vpSL) + (1 x 0.03 x vp_LL) + (5 x0.03 x vpNL) + (10 x 0.03 x vpRL) + (0.1 x 0.05 x vpSC) + (1 x 0.05 x vpLC) + (5 x 0.05 x vpNC) + (10 x 0.05 x vpRC) + (0.1 x 0.1 x vpSB) + (1 x 0.1 x vpLB) + (5 x 0.1x vpNB) + (10 x 0.1 x vpRB)) /12) [12]

Where:

vpSL = site scope with low intensity;

vpLL = local scope with low intensity;

vpNL = neighborhood scope with low intensity;

vpRL = regional scope with low intensity;

vpSC = site scope with uncomfortable intensity;

vpLC = local scope with uncomfortable intensity;

vpNC = neighborhood scope with uncomfortable intensity;

vpRC = regional scope with uncomfortable intensity;

vpSB = site scope with unbearable intensity;

vpLB = local scope with unbearable intensity;

vpNB = neighborhood scope with unbearable intensity;

vpRB = regional scope with unbearable intensity.

Pesticide Risk

This performance index considers the tendency to increase, maintain or decrease the use of pesticides in terms of frequency of use, variety of pesticides and toxicity level, over a period of three to five years. Frequency of use refers to the number of pesticide applications per year; variety signifies the number of different active ingredients used in pesticide applications; and toxicity refers to the effect on humans and other life forms who might come into contact with the pesticide.

The model defines the farm as sustainable only when at least one of processes decreases, and none increase.

The Pesticide Risk (Apr) is calculated using the expression:

$$\text{Aprvar} = 6 \times [(\text{prIF} - \text{prDF}) + (\text{prIV} - \text{prDV}) + (\text{prIT} - \text{prDT})]]$$

$$\text{Apr} = 1 / (1 + 0.41^{(0.3 \times \text{Aprvar})})$$

Where prIF and prDF specify the tendency to increase and decrease the frequency of pesticide application, respectively; prIV and prDV are the tendency to increase and decrease the variety of pesticides, respectively; and prIT and prDT represent the tendency to increase or decrease toxicity, respectively. Values will be one or zero. A value of one is assigned when the process is true and zero when the process is false. If the tendency is to maintain the process at the same intensity, the value will be zero for increasing as well as decreasing possibilities.

Landscape Dimension

The Landscape Dimension index (LDim) measures the management and conservation practices carried out on any portion of the farm, such as natural habitats, permanent preservation areas and wildlife corridors. Other processes measured include the condition of areas of geotechnical risk, practices associated with fire management and agricultural management of land use units on the farm. An unsustainable condition will be reached when landscape management or conservation practices are inappropriate to the conversation of natural resources.

Landscape Dimension (LDim) incorporates the arithmetic mean of at least five of the following performance indices: Natural Habitat Conservation, Permanent Protection Area Conservation, Wildlife Corridor Conservation, Management Condition, Geotechnical Risk and Fire Risk (Figure 5).

The modeling of these performance indices requires calculation of the territorial distribution of the different land uses inside the farm.

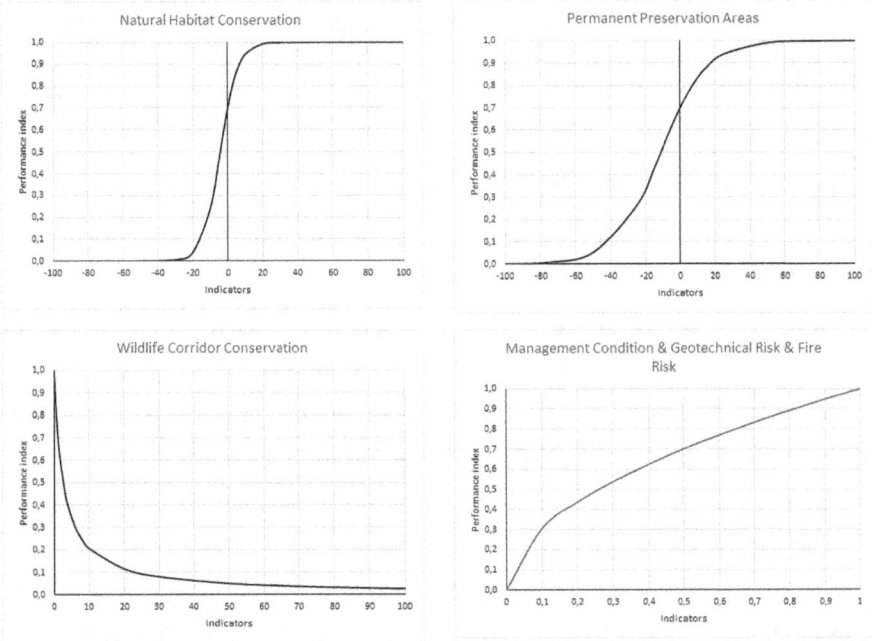

Figure 3.6: Logistic model to compute performance indices for the Landscape Dimension.

45

Natural Habitat Conservation

This performance index measures the condition of natural habitats on the farm and conservation practices carried out there. Within the farm, this index checks what type of conservation management is performed within natural habitat areas. Only areas destined for natural habitat conservation are considered in this calculation and areas of permanent preservation inside the farm are excluded.

This index also takes into account the legal limit required for the existence of natural resource conservation areas. The model considers sustainable forest management practices a better land use for these areas; abandoned areas or land uses with higher intensity of natural resource exploration will be unsustainable.

The Natural Habitat Conservation (Lnh) performance index is calculated as follows:

Lnhvar = = 100 x (((0.7 x nhNW) + nhNS + (0.8 x nhNE) + (0.6 x nhNP) + (0.4 x nhVS) + (0.2 x nhNR) + (0.1 x nhSAF)) / (nhNW + nhNS + nhNE + nhNP + nhVS + nhNR + nhSAF + nhAG))

Lnh = 1 / (1+ 0.43 x (2.71828 $^{(-0.2 \text{ x Lnhvar})}$))

Where:

nhNW = areas under natural vegetation without any management;

nhNS = areas under natural vegetation managed using sustainable forest management practices;

nhNE = areas under natural vegetation managed for wood without sustainable forest management practices;

nhNP = areas under natural vegetation managed for non-wood products without sustainable forest management practices;

nhVS = areas under secondary natural vegetation with any type of management;

nhNR = areas at the beginning of the recovery process of regeneration of the natural vegetation with any type of management;

nhSAF = areas under use as agroforestry systems; and

nhAG = areas under any agricultural use.

All indicators are reported in hectares.

Permanent Preservation Areas

This performance index measures the state of conservation of permanent preservation areas inside the farm, excluding areas that were included in natural habitat conservation. Although some anthropogenic pressures may be present, the permanent conservation areas are sustainable only if pressure from human activities has not affected the natural distribution and occurrence of species in the area.

The Permanent Preservation Areas (Lpp) performance index is calculated using the expression:

$$Lppvar = 100 \times ((ppNC + (0.3 \times ppVP) + (0.1 \times ppSF)) / (ppNC + ppVP + ppSF + ppAG))$$

$$Lpp = 1 / (1 + 0.43 \times (2.71828^{(-0.075 \times Lppvar)}))$$

Where:

$ppNC$ = permanent preservation area that keeps its physiognomy and diversity of biological species intact in relation to its natural condition;

$ppVP$ = permanent preservation areas that have suffered anthropogenic impacts with changes in species richness or diversity, but maintain the natural physiognomy of their vegetation;

$ppSF$ = permanent preservation areas strongly impacted by anthropogenic pressures and transformed into secondary forest or abandoned areas and;

$ppAG$ = permanent preservation areas that have any other land use.

All indicators are reported in hectares.

Wildlife Corridor Conservation

This performance index measures the state of conservation of wildlife corridors and verifies the management applied to areas on the farm that provide passage for wildlife. It also evaluates territorial discontinuity and fragmentation of wildlife corridors. The sustainable condition is attained when fragmentation is absent, which is measured in relation to the number of fragments and percentage of fragmented wildlife corridors.

The Wildlife Corridors Conservation (Lwc) performance index is calculated by the expression:

$$Lwc = 0.00137 / (1 - (0.998628 \times (2.71828^{(-0.00053 \times ((wcPE / 100) \times wcFR))})))$$

Where wcPE is the percentage of wildlife corridors fragmented inside the farm and wcFR is the number of fragmented wildlife corridors. The wcPE units are in hectares and wcFR is represented by counts.

Management Condition

The Management Condition performance index measures the intensity and quality of productive areas inside the farm. Sustainability depends on the proportion of farm areas where land use intensity remains below it's the land's potential to support agriculture and depends on the quality of agricultural management in these areas, where quality implies fitness of agricultural practices for each crop or land use type.

This performance index assigns an indicator to each land use unit that is available by comparing the potential land use with its current use, and requires an assessment that is described in the methods "Adequacy of Recommended Land Use - ARL". This indicator is the ARL index.

The Management Condition (Lmc) performance index is calculated by the expression:

$$Lmc = ((\sum mcGi \times mcQi \times ((mcARLi/5+1)))/2)^{0.5145}$$

Where mcGi is the fraction of the entire farm area under cultivation represented by each field, mcQi is the quality of management applied to each field and mcARLi is the adequateness of land use for each field (ARL index). For mcQi, 1 represents excellent management conditions, 0.7 good management conditions, 0.35 regular management conditions and 0 bad management conditions, for each "i" land use. The sum of mcGi should be 1, equivalent to all areas inside the farm used for agriculture. The units are proportional numbers for mcGi and mcQi and natural numbers for mcARLi.

Geotechnical Risk

This index measures the vulnerability of land use units to geotechnical processes . This vulnerability is any process, situation or event in the geological environment, occurring naturally or induced by human action, that generates social or economic damage.

Like the Management Condition performance index, this performance index also uses the adequateness of land use (ARL index) indicator.

The Geotechnical Risk (Lgr) performance index is calculated by the expression:

$$Lgr = ((\textstyle\sum mcGi \text{ x } mc\_Ti \text{ x } ((mcARLi/5+1))) / 2)^{0.5145}$$

Where mcGi is the fraction of the entire farm area under cultivation represented by each field, mcTi is the factor associated with geotechnical vulnerability and mcARLi is the adequacy of land use for each field (ARL index). The mcTi will be 1 for absence of risk, 0.8 for flood risk, 0.6 for siltation risk, 0.5 for gully erosion risk, 0.4 for subsidence risk, 0.2 for sliding risk, 0.1 for aggradation risk and 0 for landslide within the "i" land use unit. The sum of mcGi should be 1, equivalent to all of the areas inside the farm under agricultural use. The units are proportional numbers for mcGi and mcTi and natural numbers for mcARLi.

Fire Risk

The fire risk index refers to precautions that are taken to prevent or reduce natural or man-made fires that may result in injury or property damage. This performance index is associated with the reduction of flammable materials and elimination of fire use practices inside the land use unit.

Similar to the Management Condition performance index, this performance index also uses the adequateness of land use (ARL index) indicator.

The Fire Risk (Lfr) performance index is calculated by the expression:

$$Lfr = ((\textstyle\sum mcGi \times mcFi \times ((mcARLi/5+1))) / 2)^{0.5145}$$

Where mcGi is the fraction of the entire farm area under cultivation represented by each field, mcFi is the factor associated with fire risk and mcARLi is the adequacy of land use for each field (ARL index). The mcFi will be 1 for the adoption of practices that reduce or eliminate the intentional use of fire. It will be 0.85 for reducing or eliminating the amount of flammable materials; 0.7 for no change in management of the land use unit; 0.35 for adopting practices that result in an increase or accumulation of flammable materials and 0 for adoption of practices that result in increased use of fire within each "i" land use unit. The sum of mcGi should be 1, equivalent to all of the areas under agricultural use inside the farm. The units are proportional numbers for mcGi and mcFi and natural numbers for mcARLi.

Farm Environmental Quality Performance Index

In summary, the Farm Environmental Quality (FEQ) performance index is a comprehensive index that includes five dimensions, and each dimension represents a different number of processes.

The principal rule is that each process needs to be expressed along a scale of 0 to 1, independent of the type of information gathered (such as binomial numbers, natural numbers, fractions, area units, time units). Any mathematical model can be used to accomplish this.

The second rule is that the value of 0.7 represents the limit between an unsustainable and a sustainable condition.

Therefore, this system qualifies as open because new processes or indicators can be incorporated within any of the environmental dimensions, while other processes or indicators, cited above, can be removed.

The Farm Environmental Quality index will be presented on a scale of 0 to 1, and environmental dimensions or processes can easily be identified as unsustainable or sustainable.

Finally, although the FEQ is determined at the farm scale, it is possible to calculate an aggregate performance index- Regional Environmental Quality (REQ) at the watershed, county or regional scale.

$$REQ = (0.75 \times \sum FEQ\_area\_i/A) + (025 \times \sum FEQ\_unit\_i/U)$$

Where:

i = nth farm between U farms.

REQ = Regional Environmental Quality aggregated to watershed, county or regional scale;

FEQ_area_i = Territorial extent of each farm "i" multiplied by its corresponding FEQ value;

FEQ_unit_i = FEQ value for "i" farm;

A = total area of the watershed, county or region;

U = total number of farms inside watershed, county or region.

4 RECOMMENDATION FOR SUSTAINABLE LAND USE

"Someone somewhere said: if you have three soil experts analyzing a single soil sample, you will get four or more different classifications for that same soil". If this is true, how many possible interpretations can there be for the same soil? If the response is more than one, then soil classification and interpretation is not a science but a religion."

Land capability classifications are systems used for grouping soils primarily on the basis of their ability to support the production of commonly cultivated crops and pasture vegetation without deteriorating over the long term (UDSA, 2013). In many countries and regions, a better understanding of the inherent land capabilities is needed (Macaulay, 2013); however, each country would benefit from a unique classification system that reflects the particularities of its own agricultural traditions, technological level and/or environmental legislation.

The United States of America (U.S.) probably has the oldest land capability classification system in the world. The U.S. Department of Agriculture initially developed the system in the 1940s as a farm planning and soil conservation tool, but today, the system is also used to formulate other policies, like the National Resource Inventory and the Farmland Protection Policy Act, and it informs and directs the activities of many Natural Resources Conservation Service field offices. Additionally, the USDA Farm Service Agency uses the land capability classification system to identify vulnerable lands and determine appropriate management practices (UDSA, 2013).

Land capability systems are continuously being updated, sometimes involving significant modifications, like the Land Use Capability of Scotland, which was simplified from seven main categories in the 1960s to just four main categories in the 1980s (Macaulay, 2013).

On the other hand, the land capability classification system used in the US includes eight land capability categories ranked on a scale of 1 (few limitations to agriculture) to 8 (unsuitable for agriculture). Under good management, soils from classes 1 to 4 are capable of producing common field crops and pasture vegetation without reducing the soil's long-term productivity. Soils in classes 5 to 8 have limited value for commercial plant production but may be suitable for use as pasture or forestland, as well as providing opportunities for recreation, wildlife and water supply (UDSA, 2013). Canada's land capability system employs a 10-class ranking for cultivated land based on soil productivity as determined by crop yields (Manitoba Agriculture, Food and Rural Initiatives, 2013).

In Brazil, São Paulo State -- the economic center of the country – uses a land use capability system similar to that adopted by the US (Lepsch, 1991). However, other Brazilian states use a system that more resembles FAO's classification system developed in the 1960s called "System for the Evaluation of Land Agricultural Capability" (Ramalho and Beek, 1995). The Brazilian system is interesting because its land capability evaluation considers three technological levels: low, medium and high use of modern technologies for land management. The weakness of this system, however, is that it confuses modernity with increased use of mechanization. This may have been common sense during the 1970s and 1980s, when the system was developed, but it is not always true today.

The Brazilian "System for the Evaluation of Land Agricultural Capability" has six land capability categories that vary along a scale of more intensive land use to more extensive land use: annual crops, perennial crops, pastures crops, grassland, forest crops and non-agricultural areas. Each land use is then classified as good, regular, restricted or inadequate and assigned to one of three technological levels: traditional (low cultural level and low technology use), intermediate (medium cultural level and medium technology use) and modern (high cultural level and high technology use). The most intensive land use associated with the best classification at each technological level becomes the recommended land use for the corresponding area.

All capability system are interpretation of a land classification system. In other words, a land classification must precede the evaluation of land capability. Land classifications differ from one country to the next, as some countries mainly base their classifications on soil properties and soil characteristics, while others focus more on the genetic processes of soil formation.

Another weakness is that soil classification is not an exact science, such that the same soil landscape has different soil classifications due to different interpretations of the relevance of certain processes or soil properties in a given environment. Consequently, soil experts are constantly improving each country's soil classification system (Anjos *et al.*, 2013, USDA, 2013).

In addition, detailed soil survey data is not always available. The lack of detailed soil survey data limits the evaluation of land capability, and this limitation oftentimes occurs in countries where sustainable land management is urgently needed. If an inadequate scale is used for the soil survey, the land capability system might overlook soil types and soil variability across the landscape, resulting in an inappropriate land use indication and serious soil degradation risks.

In many countries, soil surveys performed at the adequate scale are difficult to come by. They can be very expensive due to the need for a large number of soil classification experts, excavation of the soil profile and many costly chemical and physical soil tests.

Finally, some types of land classification systems are defined on a conceptual basis, making their interpretation inconsistent when different soil experts make the interpretation (Delarmelinda *et al.*, 2011).

To improve land capacity classification systems, we propose a reformulation of the "System for the Evaluation of Land Agricultural Capacity" (Ramalho and Beek, 1995) by following four steps:

> a) remove the need for a soil survey (although it can still be used where this information is available);

b) adjust the system to the appropriate farm and/ or crop field scale;

c) adopt a wide conceptual review that will make the system appropriate under different modern production systems, from organic farming to plantation use and;

d) Use inexpensive landscape and soil indicators that are easy to measure using routine soil testing laboratories.

Definitions

Recommendations for Sustainable Land Use are built upon a logical framework arranged by technological level (TL), land use intensity group(IG) and land use quality class (QC).

Technological levels

The technological levels represent three different production systems.

Each production system is defined by distinct social, economic and environmental conditions, and each applies different amounts of inputs and varies in demand for financial capital.

The first technological level (1TL) consists of all production systems that are largely independent of external farm inputs and enjoy maximum utilization of internal resources. This level of technology requires low resource utilization and is typical of organic farms and ecological agriculture systems. The production system associated with 1TL depends very little on financial capital to support its main activities.

The second technological level (2TL) refers to production systems that demand medium to high intensity of external inputs to maintain the farm, but do not depend on large-scale application of resources. Examples include mineral fertilization or crops derived from plant breeding programs. Although this system depends on a greater amount of financial capital than 1TL, these funds can be amortized within one or more agricultural seasons. The demand for investments are constant with land unit area.

The third technological level (3TL) describes production systems that demand medium to high intensity of external farm inputs and depend on large-scale application of resources– for example, agricultural systems that use heavy mechanization (farm equipment). Under this production system, financial capital amortization is greater and faster when larger lands are cultivated.

In defining technological level, we assume that the technologies associated with a particular level are accessible to the farmer and he uses them to (a) surpass limitations of land use or (b) mitigate or prevent land degradation.

Intensity Groups

Intensity Groups (IG) assign potential land uses for a plot of land and range from more to less intense use: (A) annual crops, (B) perennial crops, (C) agroforestry systems, (D) grassland, (E) forest plantation and (F) natural forestry. Restrictions based in legislation are not considered in this grouping, because we assumed that any legal restriction can be applied independent of land use capability.

IG classes represent land uses recommended for the majority of farmers within each technological level and refer to the majority of agricultural crops or agricultural land, without considering specific requirements for each cultivated species.

The IGs are:

Annual Crops: high intensity land use common to annually cultivated lands, resulting in greater strain on land vulnerabilities, mainly because of high frequency use of farm equipment for activities such as soil tillage, planting, cultivation and harvesting. Pastures systems are considered annual crops if they are used for only a fraction of time over one or more decades. This group adopts the notation 1A, 2A and 3A for land use inside 1TL, 2TL and 3TL, respectively.

Perennial Crops: medium intensity land use commonly associated with minimal soil tillage or annual planting, representing perennial species or a multi-year production cycle, excluding agroforestry systems and forest plantations. This group adopts the notation 1B, 2B and 3B for land use inside 1TL, 2TL and 3TL, respectively.

Grassland: low intensity land use, where pasture grasses or leguminous species permanently cover the soil. Pastures with trees are also included in this category. This group adopts the notation 1C, 2C and 3C for land use inside 1TL, 2TL and 3TL, respectively.

Agroforestry Systems: medium intensity land use associated with a crop consortium, where at least one of the components is a tree. We expect the combination of different crops to decrease soil loss and soil degradation by erosion or nutrient exportation. This group adopts the notation 1D, 2D and 3D for land use inside 1TL, 2TL and 3TL, respectively.

Forest Plantation: low intensity land use associated with forest plantations that undergo operations such as soil tillage, planting and harvesting at wide intervals, such that low intervention results in diminished exposure to soil degradation processes. This group adopts the notation 1E, 2E and 3E for land use inside 1TL, 2TL and 3TL, respectively.

Natural Forestry: very low intensity land use, where the original vegetation is exploited without changes to the original land cover, and little to no increase of soil exposure to degradation processes occurs. This group adopts the notation 1F, 2F and 3F for land use inside 1TL, 2TL and 3TL, respectively.

Permanent preservation areas or areas with restricted land use due to legislation are not considered as an alternative to land use, and therefore, are not included in this system.

Quality Classes

Quality Classes (QC) for land use indicate the ability of the land to support particular Intensity Groups against a set of Limiting Factors for crop production within each technological level.

In practical terms, these classes assess the impact of five Limiting Factors (fertility deficiency, water deficiency, oxygen deficiency, erosion susceptibility and impediments to mechanization) on the biological and economic viability of a land use.

Biological viability refers to the ability of the land to maintain high photosynthetic production, optimizing biomass production. Economic viability refers to the condition of a certain plot of land to sustain crop production without requiring increased input levels.

The QC classes are defined as Good, Moderate, Restricted or Inapt.

Good: no single factor reduces this land use capacity to lower than 90%, in terms of its economic or biological viability. Restrictions are minimal and cause no significant reductions in productivity or benefits, and for each technological level, increases in inputs are only required in minimum quantities to replace losses due to export through harvest. The notation for this class is written in capital letters (A through F) next to each land use Intensity Group.

Moderate: limiting factors reduce the land's economic viability to below 90% but without losses in biological viability. In this class, lands require greater resource investment than lands in the Good class. Although attractive, these lands are significantly inferior to Good lands. The notation for this class is written in lowercase letters (a through f) within each land use Intensity Group.

Restricted: limiting factors reduce both the economic and biological viability of this class to below 90%. In this case, the production system requires increasing inputs for the maintenance of economic and biological sustainability. Limitations reduce productivity or benefits and require more input investments , so that costs become only marginally justified. The notation for this class is written as lowercase letters in brackets ((a) through (f)) beside each land use Intensity Group.

Inapt: limiting factors result in total loss of the land's economic and biological viability. These lands suffer increased biological degradation and an ever increasing need for inputs. Lands under these conditions have no potential for sustainable land use. For notation, 'inapt' is written beside each land use Intensity Group.

Limiting Factors

Limiting Factors (LF) represent landscape and edaphic processes that express the vulnerability of the land to certain types of use or the land's inability to supply a particular function within the landscape or production unit.

Limiting Factors include Fertility Deficiency, Water Deficiency, Oxygen Deficiency, Erosion Susceptibility and Impediments to Mechanization, and are evaluated through soil or landscape indicators.

Each Limiting Factor is associated with five degrees of limitation: null (N), light (L), medium (M), strong (S) and very strong (V) and is assigned a number that indicates the main limiting processes. These processes affect economic and/or biological viability.

The number associated with each degree of limitation describes the principal cause of limitation linked to the landscape or edaphic process, facilitating diagnosis and recommendations to mitigate or reorganize the land use inside each technological level.

Soil and Landscape Indicators

Soil and Landscape Indicators are based on landscape properties (slope, soil depth, drainage, rainfall, soil rockiness or stones on the soil) and tested soil properties (exchange cations, texture, and acidity).

To test for soil properties, soil samples are collected with an auger tool at a central point in the land use unit at depths of 0 to 25 cm (upper layer), 26 to 60 cm (middle layer) and 61-100 cm (bottom layer).

Soil samples are then dried and passed through a 2 mm sieve. Chemical analyses are carried out for calcium, magnesium and exchangeable aluminum, available potassium and sodium, remaining phosphorus, potential acidity and organic carbon. Separate soil samples are used to test for physical characteristics of soil texture and soil particle density. Upper layer soil samples are used to measure electrical conductivity and sodium adsorption ratio in saturation extract.

Soil properties like soil air capability, soil water capability and soil erodibility are estimated using pedotransfer functions.

Indicators include:

Electrical Conductivity of Soils (EC): determined from the upper soil sample using saturation extract (Embrapa, 1997). This indicator is unnecessary if no evidence of soil salinity or soil sodicity is present. The unit of measurement for EC is dS cm^{-1}.

Sodium Adsorption Ratio (SAR): sodium adsorption ratio (SAR) measures soil sodicity, as determined from analysis of water extracted from the soil (Embrapa, 1997). This indicator is unnecessary if no evidence of soil sodicity is present.

The sodium adsorption ratio is:

$$SAR = [Na^+] / \{([Ca^{2+}] + [Mg^{2+}]) / 2\}^{1/2}.$$

Where sodium, calcium, and magnesium represent the concentration of these ions in soil extract saturation in cmol L^{-1} of soil. The unit of measurement for SAR is %.

Soil Texture (Texture): distribution of sand, silt and clay in the soil sample. The content of sand, silt and clay are determined by Embrapa (1997). Soil textures are defined as:

a) Very clayey soils - clay content greater than 600 g kg^{-1}.

b) Clayey soils - clay content greater than 350 g kg^{-1} but less than or equal to 600 g kg^{-1}.

c) Silty soils - clay content less than or equal to 350 g kg^{-1} and sand content less than or equal to 150 g kg^{-1}.

d) Sandy soils – clay content $\leq$ 350 g kg^{-1} and sand content > 700 g kg^{-1}. If clay content < 150 g kg^{-1} , then silt content < 300 g kg^{-1} and the sum of proportions of sand and clay > 700 g kg^{-1} and the sum of the proportions of sand and silt > 700 g kg^{-1} and the sum of clay and silt < 300 g kg^{-1}.

e) Loam soils: all other cases that do not fit the rules described above.

Percent Base Saturation (PBS): proportion of all soil exchange sites occupied by basic cations. Soil exchange sites refer to the total amount of positively charged elements that a soil can hold: basic cations and acidic cations.

Basic cations are calcium (Ca^{+2}), magnesium (Mg^{+2}), potassium (K^{+1}) and sodium (Na^{+1}) on exchangeable sites in the soil. The acidic cations are hydrogen (H^{+1}) and aluminum (Al^{+3}) on exchangeable sites in the soil. The unit of measurement for cations in the soil is cmol L^{-1} (Embrapa, 1997).)

The soil layer is considered dystrophic if PBS < 50% and eutrophic if PBS $\geq$ 50% (Embrapa, 2006).

<u>Soil Basic Cations Sum</u> (SB): amount of soil basic cations in the soil calculated by:

SB = (SBupper x 0.6) + (SBmiddle x 0.32) + (SBbottom x 0.08),

Where, SBupper, SBmiddle and SBbottom are the basic cations in upper, middle and bottom layers of the soil sample, respectively, determined by Embrapa (1997) in cmol L^{-1}.

<u>Soil Aluminum Saturation</u> (SAS): percent of exchangeable aluminum in the soil defined as the sum of exchangeable soil sites occupied by basic cations and aluminum cations, both determined by Embrapa (1997), in cmol L^{-1}.

<u>Soil Carbon</u> (SC): carbon held within the soil, primarily in association with its organic content. Soil carbon is primarily composed of biomass and non-biomass sources. If measured as total soil carbon, the value must be multiplied by a factor of 0.7. The unit of measurement for SC is dag L^{-1} (Embrapa, 1997).

<u>Cations Exchange Capacity</u> (CEC): sum of the basic cations and acidic cations, in cmol L^{-1} (Embrapa, 1997).

<u>Clay</u>, <u>silt</u> and <u>sand</u> <u>content</u>: determined from soil dispersion with sodium hydroxide or other dispersant with similar power. Silt can be determined by calculating the difference between 100% and clay plus sand percentages, or by weighing (silt + clay) and subsequently removing the mass of clay from the sample (Embrapa, 1997). The unit of measurement is dag kg^{-1}.

<u>Clay Activity</u> (CA): the value of the CEC divided by the percent Clay content present in soil sample. (Embrapa, 1997). The unit of measurement for CA is cmol L^{-1}.

Remaining Phosphorus (RP): a measure of the soil's ability to maintain a certain level of P in solution and determined by the equilibrium concentration of phosphorus after the addition of 60 mg L^{-1} of P (Alvarez *et al..*, 2000). The unit of measurement for RP is mg L^{-1}.

Effective Soil Depth (DEPTH): presence or absence of physical impediments to root penetration up to depths of 100 cm. If any impediment exists, the depth at which the impairment occurs is recorded. The unit of measurement for DEPTH is cm.

Annual Water Inputs (AWI): sum of the average annual rainfall, defined according to the region's climatic data plus any addition of irrigation water. The unit of measurement for AWI is mm $year^{-1}$.

Dry Season (DRY): number of months per year during which total evapotranspiration is greater than rainfall . The unit of measurement for DRY is an integer.

Soil Water Available (SWA): estimate of the storage capacity of water in the soil up to a depth of the 100 cm, expressed in cm of available water, by the expression:

SWA = + aw1 (aw2 x 0.78) + (aw3 x 0.50)

Where aw1, aw2 and aw3 are the amounts of available water stored in soil layers 0 cm to 25 cm, 26 to 60 and 61 cm to 100 cm, respectively.

In each layer, available water (aw) is estimated by (Arruda *et al,* 1987):

SWA = (FC - PWP) / 10 x T x SD

Where:

FC = water in field capacity (%), estimated by the equation: {3.07439 + 0.629239 x (100 - sand) + 0.00343813 x (100 - sand) ^{2}};

PWP = permanent wilting point (%), estimated by the equation: {398.889 x (100 - sand) / 1308.09 + (100 - sand)};

Sand = sand content in each soil layer;

T = thickness of the sampling layer 25, 35 and 40 cm for soil depth layers 0 to 25 cm, 26 to 60 and 61 to 100 cm, respectively;

SD = soil density, in dag kg^{-1}, calculated by (Benites et al., 2007): SD = 1.5600 - (1.0005 x clay) - (0.010 x SC) + (0.0075 x SB);

Clay = clay content in each soil layer, in dag kg^{-1};

SC = soil carbon in each soil layer, in dag L^1;

SB= basic cations in each soil layer determined by Embrapa (1997), in cmol L^{-1}.

Slope (SLP): the average land declivity (%) measured in the field, using a clinometer.

Floodplains (FP): lands that represent sedimentation areas in river basins, lands that receive surface water from higher points in the watershed or lands with periodic flooding due to overflowing drainage systems. The land must exhibit plain relief (SLP < 3 %).

Highland (HL): an elevated region or plateau.

Upland (UL): land elevated above other land or land where flooding never occurs.

Erodibility (EDL): estimated by Williams et al.'s equation (Song et al., 2005), which only requires variables that are easy to measure:

EDL = SN1 x SN2 x SN3 x SN4

Where:

EDL = erodibility in $t.h.MJ^{-1}$ mm;

SN1: 0.2 + 0.3 exp(-0.0256 x sand x (1 - silt/100))

SN2: $(silt / (clay + silt))^{0.3}$

SN3: 1 - (0.25 x SC / (SC + exp (3.72 - 2.95 x SC)))

SN4: 1 - ((0.7 x SN5) / (SN5 + exp (-5.51 + 22.9 x SN5))

SN5: 1 - (sand/100).

Silt = silt content in upper layer, in dag kg^{-1};

Sand = sand content in upper layer, in dag kg^{-1};

Clay = clay content in upper layer, in dag kg^{-1};

SC = soil carbon in upper layer, in dag L^{-1}.

Degree of Soil Oxygenation (SOx): estimate of restrictions due to soil drainage conditions that constrain adequate development of the radicular system of crops adapted to highland conditions and ranked from 0 to 3. This limiting factor depends on the depth that restrictions occur indicating drainage deficiency, such as elevated groundwater or gray color in the soil.

Characteristics related to drainage restrictions are assessed at the time of soil sampling.

Soil oxygenation is ranked by degree:

- Degree 3: if soil depth is less than 100 cm, it displays physical constraint in the upper layer, or it has gray color in the upper or medium layer;

- Degree 2: if lower soil layer is gray in color or it displays mottled colors in the medium or upper layer;

- Degree 1: if the soil exhibits mottled colors in the bottom layer; and

- Degree 0: if no impediment to soil oxygenation.

Mechanization Constraints (MC): represents impediments to the operation of farm equipment associated with soil rockiness and stoniness and ranked with scores ranging from 0 to 2.

Rockiness represents rocks present within the soil and at a depths that negatively affect farm equipment operation.

Stoniness represents the presence of stones and rocks on the surface of the soil.

Degrees of MC:

- Degree 2: when rockiness exceeds 25% or stoniness is more than 30%;

- Degree 1: rockiness present but not exceeding 25% and stoniness present but not exceeding 30%; and

- Degree 0: no rockiness or stoniness in the soil.

Degrees of Limitation

The Degrees of Limitation - null (N), light (L), medium (M), strong (S) and very strong (V) - are defined for each Limiting Factor (Fertility Deficiency, Water Deficiency, Oxygen Deficiency, Erosion Susceptibility and Impediments to Mechanization).

Fertility Deficiency

Fertility Deficiency indicates soil chemical conditions that can limit crop development, such as the presence of toxic elements and low nutrient reserves. These restrictions will be more important at the first technological level, since at the second and third technological levels, agricultural practices can mitigate these problems.

Null Limitation: Lands where the soils have high nutrient reserves without salinity or sodicity, soils with no acidity in the upper layer, or acidic soils with high nutrient reserves associated with high activity clays and medium to low adsorption capacity of phosphate at moderately deep or deeper soil layers.

The notation adopted for this class is N or f_N.

The decision rules for soils with Null Limitation are:

Soils with non-sandy texture in the upper or middle layer, with SB greater than 6 cmol dm^{-3} and low salinity (EC less than or equal to 4 dS m^{-1} and SAR less than or equal to 6 %), and that meet one of the following conditions:

1 - When associated with low activity clays (CA less than or equal to 27 cmol dm^{-3}) in the middle layer, soils must be eutrophic (PBS greater than 50%) in the upper and middle layers; or

2 - When associated with high activity clays (CA greater than 27 cmol dm^{-3}) in the middle layer, soils must provide medium to low adsorption capacity of phosphate (RP greater than 5 mg L^{-1}) in the middle layer and have DEPTH greater than 50 cm.

Light Limitation: Lands where soils have medium to high nutrient reserves and no salinity or sodicity, unless weak acidity. Distinguished from soils with Null Limitation by their high adsorption capacity for phosphate, shallow soil depth (DEPTH less than or equal to 50 cm), or dystrophic soil characteristics (PBS less than or equal to 50 %).

The notation adopted for this class is L or f_L.

The Light Limitation decision rules are:

Soils with non-sandy texture in the upper or middle layer, with SB greater than 6 cmol dm^{-3} and low salinity (EC less than or equal to 4 dS m^{-1} and SAR less than or equal to 6 %), and that meet one of the following conditions:

1- When associated with low activity clays (CEC less than or equal to 27 cmol dm^{-3}) in the middle layer, soils must be eutrophic (PBS greater than 50%) in the upper and middle layers and display SAS less than or equal to 30% in the middle layer (limiting factor # 11 or f_L/a); or

2 - When associated with high activity clays (CEC greater than 27 cmol dm^{-3}) in the middle layer, soils must be shallow (DEPTH less than or equal to 50 cm) or have high adsorption capacity for phosphate in the middle layer (RP $\leq$ 5 mg L^{-1}) (limiting factor # 10 or f_L/n).

Medium Limitation: Lands where soils have moderate salinity or sodicity, or lands where the soil has low to medium nutrient reserves associated with moderate to strong acidity.

The notation adopted for this class is M or f_M.

The decision rules for Medium Limitation are:

Weakly saline soils (4 dS m^{-1} less than or equal to EC less than or equal to 8 dS m^{-1}) or weakly sodic soils (6% < SAR <15%) (Limiting factor # 3 or f_M/s).

If soils are not saline or non-sodic, then:

1 - Soils with SB greater than 6 cmol L^{-1} must have high aluminum saturation in the middle layer (SAS greater than 30 %) (limiting factor # 8 or f_M/a);

2 – Soils with SB between 3 cmol dm ^{3} and 6 cmol dm^{-3} must display PBS less than or equal to 50 % in upper or middle layer and have SAS less than or equal to 50% in middle layer %) (limiting factor # 9 or f_M/n).

Strong Limitation: Lands where soils have medium to low salinity and are sodic, or when not saline or non-sodic, soils with low nutrient reserves but eutrophic or dystrophic characteristics when aluminum saturation is low.

The notation adopted for this class is S or f_S.

The decision rules for Strong Limitation are:

All soils with high sodicity (SAR greater than 15%) or with medium salinity (8 dS m^{-1} less than or equal to EC 15 dS m^{-1}) (limiting factor # 2 or f_S/s:).

Otherwise, if soils with non-sandy texture in the upper or middle layer and SB between (1.5 cmol dm $^{-3}$ and 3.0 cmol dm^{-3}), then:

1 – If eutrophic in the upper and middle layer (limiting factor # 7 or f_S/n); or

2 – If dystrophic in either the upper or middle layer, but having SAS less than or equal to 30% (limiting factor # 6 or f_S/a).

Very Strong Limitation: Lands where soils display high salinity, are extremely poor in nutrients, or have low nutrient reserves associated with high acidity.

The notation adopted for this class is V or f_V.

The Very Strong Limitation decision rules are:

All soils with CE greater than 15 dS m^{-1} (limiting factor # 1 or f_V/s).

Otherwise, all soils with sandy texture in the upper and middle layers or with low nutrient reserves in the soil profile (SB less than or equal to 1.5 cmol dm^{-3}) associated with SAS less than or equal to 30% in the middle layer (limiting factor # 4 or f_V/n);

Also, soils with low nutrient reserves in the soil profile (SB less than or equal to 1.5 cmol dm^{-3}) associated with SAS greater than 30% in the middle layer, or soils with SB between 1.5 to 3.0 cmol dm^{-1}, eutrophic in the upper or middle layer (PBS less than or equal to 50%) and SAS greater than 30% in the middle layer (limiting factor # 5 or f_V/a).

The entire framework for fertility deficiency is composed of 19 decision rules (Table 4.1)

Table 4.1. Decision rules for fertility deficiency.

Step	Decision	If true	If false
1	EC > 15	#1: f_V	2
2	EC > 8	#2: f_S	3
3	SAR > 15	#2: f_S	4
4	EC > 4	#3: f_M	5
5	SAR > 6	#3: f_M	6
6	Upper Texture and middle Texture = "Sandy"	#4: f_Sn	7
7	SB ≤ 1.5	8	9
8	Middle SAS ≥ 30	#5: f_Ma	#4: f_Sn
9	SB ≤ 3.0	10	12
10	Eutrophic in upper and middle layer	#7: f_Sn	11
11	Middle SAS > 30	#5: f_Ma	#6: f_Sa
12	SB ≤ 6.0	13	15
13	Middle SAS > 30	#8: f_Ma	14
14	Eutrophic in upper and middle layer	#9: f_Mn	#10: f_La
15	Middle CA < 27	16	18
16	Eutrophic in upper and middle layer	N	17
17	Middle SAS > 30	#8: f_Ma	#10: f_La
18	Middle RP <5	#11: f_Ln	19
19	DEPTH < 51	#11: f_Ln	N

Water Deficiency

Water Deficiency occurs when climatic conditions associated with landscape and soil physical characteristics limit water availability for normal plant development (Figure 2). This interpretation is mostly applicable to agricultural systems that use no irrigation, although it can also be applied to irrigated areas if total water input depends on irrigation water coupled with rain water. Irrigation can be used at any of the Technological Levels, depending only on local conditions inside the farm.

Null Limitation: Lands with no water limitation to crop development at any time of year, where rainfall and/or irrigation associated with large soil water storage capacity adequately meets crop water demands, or floodplains associated with groundwater close to the soil surface.

The notation adopted for this class is N or w_N.

The decision rules are:

All lands located in floodplains with high groundwater and AWI greater than 1000 mm, or lands with AWI between 1000 and 1500 mm and no dry season, or lands with AWI between 1500 and 2000 mm and a dry season lasting up to one month, but never in soils that display CA greater than 27 cmol dm^{-3} in the middle layer, or lands with AWI greater than 2000 mm and a dry season lasting up to one month.

Light Limitation: Lands with slight water deficiency, where only one crop per year is feasible without losses in productivity. A second crop in the same year will result in 80% lower productivity than the first crop. Perennial crops remain unaffected as long as the dry season does not coincide with lower plant physiological activity.

The notation adopted for this class is L or w_L.

The decision rules are:

Floodplain lands with AWI between 500 and 1000 mm and SWA greater than 75 mm (limiting factor # 14 or w_L/p).

If not floodplains, then:

Lands with AWI between 500 and 1000 mm and no dry season (limiting factor # 19 or w_L/d); or

Lands with average rainfall between 1000 and 1500 mm and a dry season lasting one to two months (limiting factor # 19 or w_L/d), lands with average rainfall between 1500 and 2000 mm and a dry season longer than one month or soils with CA greater than 27 cmol dm^{-3} in the middle soil layer; or land with AWI greater than 2000 mm associated with a dry season longer than three months (limiting factor # 21 or w_L/d).

Medium Limitation: Lands with moderate water deficiency, where crop development is restricted by the lack of water for a period of four to six months. On these lands only one annual crops is viable each year. Yields of a second crop will always be less than 50% of the first. Perennial crops with no drought adaptations will be negatively affected.

The notation adopted for this class is M or w_M.

The decision rules are:

Floodplain lands with AWI between 500 -1000 mm in soils with SWA less than or equal to 75 mm (limiting factor # 15 or w_Mp).

If not floodplains, then:

Land with AWI between 500 - 1000 mm and a dry season lasting up to one or two months (limiting factor # 18 or w_Mp); or

Lands with AWI between 1000-1500 mm and a dry season longer than three months (limiting factor # 20 or w_Md).

Strong Limitation: Lands with strong water deficiency due to low soil AWI associated with a six to nine month dry season. Annual crop development is only possible during the rainy season. Perennial crops not adapted to drought conditions cannot survive in this environment.

The notation adopted for this class is S or w_S.

The decision rules are:

Floodplain lands with AWI between 250 and 500 mm (limiting factor # 13).

If not floodplains, then:

Lands with rainfall between 250 and 500 mm in soils with SWA greater than 75 mm (limiting factor # 17); or

Lands with rainfall between 500 and 1000 mm and a dry season longer than six months in soils with SWA greater than 40 mm (limiting factor # 17); or

Lands with rainfall between 500 and 1000 mm and a dry season longer than three months (limiting factor # 17); or

Very Strong Limitation: Lands with severe water stress, usually with a dry season greater than nine months. Typically, only highly specialized crops adapted to severe water stress can develop on these lands.

The notation adopted for this class is V or w_V.

The decision rules are:

Floodplain lands with AWI below 250 mm (limiting factor # 12)

If not floodplain:

Lands with AWI below 500 mm (limiting factor # 16); or

Lands with rainfall between 250 and 500 mm in soils with SWA less than or equal to 75 mm (limiting factor # 17); or

Lands with rainfall between 500 and 1000 mm and a dry season longer than six months in soils with SWA less than or equal to 40 mm (limiting factor # 16);

The entire framework for water deficiency is made up of seventeen decision rules (Table 4.2)

Table 4.2. Decision rules for water deficiency.

Step	Decision	If true	If false
1	Floodplains Area	2	6
2	AWI > 1000	N	3
3	AWI > 500	4	5
4	SWA > 75	#14: w_L	#15: w_M
5	AWI > 250	#13: w_S	#12: w_V
6	AWI > 250	7	#16: w_V
7	AWI >500	9	8
8	SWA > 75	#17: w_S	#16: w_V
9	AWI > 1000	14	10
10	Dry Season > 6	11	12
11	SWA > 40	#17: w_S	#16: w_V
12	Dry Season > 6	#17: w_S	13
13	Dry Season > 3	#18: w_M	#19: w_L
14	AWI > 1500	17	15
15	Dry Season > 1	#20: w_M	16
16	Dry Season > 0	#21: w_L	N
17	AWI > 2000	20	18
18	Dry Season > 1	#21: w_L	19
19	Middle CA >27	#21: w_L	N
20	Dry Season > 1	#21: w_L	N

Oxygen Deficiency

Oxygen deficiency describes lands where soil oxygenation restricts crop development because of deficient soil drainage. Affected crops include those whose root systems are not adapted to floodplain conditions. Current conditions for soil drainage that include man-made structures built to drain the soil should be considered instead of the land's natural condition.

Null Limitation: Lands with no oxygen limitation to crop development at any time of year.

The notation adopted for this class is N or o_N.

All lands with well drained or excessively drained soils, located in uplands with slops greater than 3% and where soil oxygenation equal to 0, or in highlands, with slopes less than 3% where soil oxygenation equal to 0 and low activity clay is present in the middle and bottom soil layers.

Light Limitation: Lands with little impairment to crops sensitive to excess water during the rainy season due to aeration conditions generally associated with moderately drained soils.

The notation adopted for this class is L or o_L.

The decision rules are:

Uplands with slopes greater than 3% and soils with soil oxygenation equal to 1 (limiting factor # 27); or

Highlands, with slopes less than 3% and soils with soil oxygenation equal to 1 or 0 associated with high activity clays in the middle or bottom layer (limiting factor # 25); or

Floodplain lands where soil oxygenation equal to 0 and middle layer soil texture is not very clayey and without high activity clays in the middle or bottom layer (limiting factor # 25).

Medium Limitation: Lands where most crops are sensitive to oxygen deficiency and fail to develop satisfactorily. These lands are associated with soils considered imperfectly drained and subject to occasional flooding.

The notation adopted for this class is M or o_M.

The decision rules are:

Uplands with slopes greater than 3% and soils with soil oxygenation greater than 1 (limiting factor # 26) or;

Uplands with slopes less than 3% in areas with soil oxygenation equal to 2 (limiting factor # 24);

Floodplain lands with soil oxygenation equal to 1 associated with very clayey soil texture in the middle layer or high activity clays in the middle or bottom layer (limiting factor # 24).

Strong Limitation: Lands with severe oxygen deficiencies, only allowing for development of adapted crops. These lands are considered poorly drained or very poorly drained soils and require intensive drainage works for cultivation to succeed . Such land is commonly subject to production losses due to floods.

The notation adopted for this class is S or o_S.

The decision rules are:

Uplands with slopes less than 3% in areas with soil oxygenation equal to 3 (limiting factor # 23); or

Floodplain lands with soil oxygenation equal to 2 (limiting factor # 23).

Very Strong Limitation: Lands with severe oxygen deficiencies, and where infrastructure designed to improve the land is prohibitively expensive and largely out of reach for farmers, unless they have access to government or union assistance.

The notation adopted for this class is V or o_V.

The decision rule is:

Floodplain lands with soil oxygenation equal to 3 (limiting factor # 22)

The entire framework for oxygen deficiency is composed of twelve decision rules (Table 4.3)

Table 4.3. Decision rules for oxygen deficiency.

Step	Decision	If true	If false
1	Floodplain Area	2	6
2	SOx = 3	#22: o_V	3
3	SOx =2	#23:o_S	4
4	SOx = 1	#24: o_M	5
5	Middle texture = "very clayey" or Middle CA > 27 or Bottom CA > 27	#24: o_M	#25: o_L
6	Slope > 3	11	7
7	SOx = 3	23:o_S	8
8	SOx = 2	#24: o_M	9
9	SOx = 1	#25: o_L	10
10	Middle CA > 27 or Bottom CA > 27	#25: o_L	N
11	SOx > 1	#26: o_M	12
12	SOx > 0	#27: o_L	N

<u>Erosion Susceptibility</u>

Erosion susceptibility represents different conditions related to risks of erosion, sedimentation and environmental harm associated with crops in each landscape. Erosion risk is evaluated by the intrinsic quality of the soil type and topographic soil characteristics in the landscape.

Null Limitation: Lands with soils unsusceptible to erosion loss. On these lands, water runoff is minimal and soils are resistant to disintegration.

The notation adopted for this class is N or e_N.

The decision rule is:

Land where slope is less than 3% with erodibility less than or equal to 0.2 mm^{-1} th MJ^{-1}, and soil oxygenation is less than or equal to 3.

Light Limitation: Lands with soils that exhibit good physical properties, and where soil erosion processes can be avoided by simple conservation practices.

The notation adopted for this class is L or e_L.

The decision rules are:

Lands with slope less than or equal to 3% and erodibility $\leq$ 0.2 mm^{-1} t h MJ^{-1}, or if erodibility greater than 0.2 mm^{-1} t h MJ^{-1}, then soil oxygenation = 3 (limiting factor # 28) or,

Lands with slope between 3% and 8% and erodibility less than or equal to 0.2 t h MJ^{-1} mm^{-1} and soil oxygenation less than or equal to 3 (limiting factor # 28).

Medium Limitation: Lands with moderate susceptibility to erosion, which produces ridges and gullies when conservation principles are neglected, and as a result require the adoption of erosion control practices.

The notation adopted for this class is M or e_M.

The decision rules are:

Lands with slope between 3% and 8% and erodibility $\geq$ 0.2 t h MJ^{-1} mm^{-1}, and where soil oxygenation equal to 3 (limiting factor # 29); or

Lands with slope between 8% and 20% and soil oxygenation less than or equal to 3 (limiting factor # 29).

Strong Limitation: Lands with severe erosion susceptibility, requiring intensive agricultural practices. Either strong slopes on these lands provoke damaging runoff, or soil physical characteristics are extremely unfavorable.

The notation adopted for this class is S or e_S.

The decision rules are:

Lands with slope between 8% and 20% and soil oxygenation = 3 (limiting factor # 30 or e_S); or

Lands with slope between 20% and 45% (limiting factor # 30)

Very Strong Limitation: Lands with severe erosion susceptibility, where erosion control practices are expensive and uneconomical. These lands occur in mountainous areas or in very undulating terrain.

The notation adopted for this class is V or e_V.

The decision rules are:

Land with slops greater than 45% (limiting factor # 31).

The entire framework for Erosion Susceptibility is made up of nine decision rules (Table 4.4)

Table 4.4. Decision rules for Erosion Susceptibility.

Step	Decision	If true	If false
1	Slope > 3	2	4
2	EDL ≤ 0.2	3	#28: e_L
3	SOx <3	N	#28: e_L
4	Slope > 8	7	5
5	EDL ≤ 0.2	6	#29: e_M
6	SOx < 3	#28: e_L	#29: e_M
7	Slope > 20	9	8
8	SOx < 2	#29: e_M	#30: e_S
9	Slope > 45	#31: e_V	#30: e_S

Impediments to Mechanization

Impediments to mechanization characterize the different conditions that limit the use of agricultural machinery. Land that has no impediments to mechanization must cover a larger area than the defined minimum size where the mechanization is economically viable. Very small areas that have no impediment to mechanization but are scattered among other areas that do have limitations to mechanization need be classified according to the more restrictive degree of the limitation.

Null Limitation: Lands that have no impediment to mechanized operations at any time of the year and whose efficiency from machinery and equipment is over 90%.

The notation adopted for this class is N or m_N.

Lands with slope less than or equal to 3 %, where MC is less than 1 and soil oxygenation is less than 2, and if the soil has a very clayey or clayey texture, the top layer contains no high activity clays.

Light Limitation: Lands where mechanization is possible at almost any time of year and whose efficiency of mechanized operations is between 75 - 90%.

The notation adopted for this class is L or m_L.

The decision rules are:

Lands with slope less than or equal to 3 %, where MC is less than 1 and soil oxygenation is less than 2, but where the soils have a very clayey or clayey texture associated with high activity clays in the top layer (limiting factor # 33); or

Lands with slope less than or equal to 3% in soils with MC equal to 2 (limiting factor # 33); or

Lands with slope between 3% and 8% and soils with MC less than 1 (limiting factor # 34).

Medium Limitation: Lands that do not ordinarily allow the use of mechanization throughout the year, and where the efficiency of mechanized operations is between 50 - 75%.

The notation adopted for this class is M or m_M.

The decision rules are:

Lands with slope less than or equal to 3 % and with MC less than 2, or if MC is less than 1, soil oxygenation must be less than 2 (limiting factor # 32 or m_M); or

Lands with slope between 3% and 8% in soils with MC equal to 1 (limiting factor # 35).

Strong Limitation: Lands where mechanization is limited to animal traction or specialized agricultural equipment, and where the efficiency of mechanized operations is less than 50%.

The notation adopted for this class is S or m_S.

The decision rules are:

Lands with slope between 3% and 8% in soils with MC greater than or equal to 2 (limiting factor # 36); or

Lands with slope between 8% and 20% in soils with MC equal to 1 (limiting factor # 36); or

Lands with slope between 20% and 45% in soils with MC less than 1 (limiting factor # 38).

Very Strong Limitation: Lands where even the use of animal traction or specialized agricultural equipment is difficult.

The notation adopted for this class is V or m_V.

The decision rules are:

Lands with slope between 20% and 45% in soils with MC greater than or equal to 1, or lands with slopes greater than 45% (limiting factor # 37).

The whole framework for Impediments to Mechanization is made up of 13 decision rules (Table 4.5).

Table 4.5. Decision rules for Impediments to Mechanization.

Step	Decision	If true	If false
1	Slope > 3	5	2
2	MC < 1	4	3
3	MC < 2	#33: m_L	#32: m_M
4	SOx<2	5	#32: m_M
5	Upper texture very clayey or clayey and upper CA >27	#33: m_L	N
6	Slope > 8	9	7
7	MC <1	#34: m_L	8
8	MC <2	#35: m_M	#36: m_S
9	Slope >20	12	10
10	MC <1	#35: m_M	11
11	MC <2	#36: m_S	#37: m_V
12	Slope > 45	#37: m_V	13
13	MC <1	#38: m_S	#37: m_V

Recommendations for Sustainable Land Use

Recommendations for Sustainable Land Use (ReSLU)are assigned separately to each technological level (1TL, 2TL or 3TL).

First, for each technological level, the Quality Class (Good, Moderate, Restricted or Inapt) is assigned to each Intensity Group (annual crops, perennial crops, agroforestry systems, grassland, plantation forestry or natural forestry) using the Degrees of Limitation (Null, Light, Medium, Strong or Very Strong) that were defined for each Limiting Factor (Fertility Deficiency, Water Deficiency, Oxygen Deficiency, Erosion Susceptibility and Impediments to Mechanization) (Tables 4.6 to 4.10).

Tables 4.6 to 4.10 explain how land characteristics at each location determine the constraints for sustainable land use.

Then, the QC identified for each Limiting Factor is transferred to the corresponding table for each technological level: 1TL (Table 4.11), 2TL (Table 4.12) and 3TL (Table 4.13). Next, the most restrictive QC for the land use in each IG is selected. This is represented by the last line on each table.

Finally, the land use Intensity Group with the least restrictive QC is selected as the recommended land use (4.14).

Tables 4.6 to 4.14 help define the recommended land use, but this can also be accomplished with computer programs using vectors.

Table 4.6 Framework for Quality Classes (QC) inside each Intensity Groups (IG) and Technological Level (TL) in relation to Fertility Deficiency Limiting Factors (LF)

LF	AC	CP	AFS	G	FP	NF
			Technological Level: 1			
f_N	A	B	C	D	E	F
f_Ln	a	b	C	D	E	F
f_La	(a)	(b)	c	D	E	F
f_Mn	(a)	(b)	c	d	e	F
f_Ma	(a)	(b)	c	d	e	F
f_Ms	inapt	inapt	(c)	d	e	F
f_Fn	inapt	inapt	(c)	d	e	F
f_Fa	inapt	inapt	(c)	d	e	F
f_Fs	inapt	inapt	inapt	(d)	(e)	F
f_MFn	inapt	inapt	inapt	(d)	(e)	F
f_MFa	inapt	inapt	inapt	(d)	(e)	F
f_MFs	inapt	inapt	inapt	inapt	inapt	F
			Technological Level: 2			
f_N	A	B	C	D	E	F
f_Ln	A	B	C	D	E	F
f_La	A	B	C	D	E	F
f_Mn	A	B	C	D	E	F
f_Ma	a	b	c	D	E	F
f_Ms	(a)	(b)	(c)	D	E	F
f_Fn	a	b	c	d	e	F

Continue…

LF	AC	CP	AFS	G	FP	NF
			Technological Level: 2			
f_Fa	(a)	(b)	(c)	D	E	F
f_Fs	inapt	inapt	inapt	d	e	F
f_MFn	(a)	(b)	(c)	(d)	(e)	F
f_MFa	(a)	(b)	(c)	d	e	F
f_MFs	inapt	inapt	inapt	(d)	(e)	F
			Technological Level: 3			
f_N	A	B	C	D	E	F
f_Ln	A	B	C	D	E	F
f_La	A	B	C	D	E	F
f_Mn	A	B	C	D	E	F
f_Ma	A	B	C	D	E	F
f_Ms	(a)	(b)	(c)	D	E	F
f_Fn	A	B	C	D	E	F
f_Fa	A	b	c	D	E	F
f_Fs	inapt	inapt	inapt	d	e	F
f_MFn	a	b	c	D	E	F
f_MFa	a	(b)	(c)	d	e	F
f_MFs	inapt	inapt	inapt	(d)	(e)	F

Where IGs are Annual Crops (AC), Perennial Crops (PC), Grassland(G), Agroforestry Systems (AFS). Forest Plantation (FP) and Natural Forestry (NF) and QC are Good (capital letters), Moderate (lowercase letters), Restricted (lowercase letters in brackets) and Inapt.

Table 4.7 Framework for Quality Classes (QC) inside each Intensity Group (IG) and Technological Level (TL) in relation to Water Deficiency Limiting Factors (LF)

LF	AC	CP	AFS	G	FP	NF
Technological Level: 1						
w_N	A	B	C	D	E	F
w_Ld	A	b	c	D	E	F
w_Lp	a	b	c	D	E	F
w_Md	a	(b)	(c)	d	e	F
w_Mp	(a)	(b)	(c)	(d)	(e)	F
w_F	(a)	inapt	inapt	(d)	(e)	f
w_MF	inapt	inapt	inapt	inapt	inapt	f
Technological Level: 2						
w_N	A	B	C	D	E	F
w_Ld	A	B	C	D	E	F
w_Lp	a	b	c	D	E	F
w_Md	a	b	c	d	e	F
w_Mp	a	(b)	(c)	d	e	F
w_F	(a)	(b)	(c)	(d)	(e)	f
w_MF	inapt	inapt	inapt	inapt	(e)	f
Technological Level: 3						
w_N	A	B	C	D	E	F
w_Ld	A	B	C	D	E	F
w_Lp	A	B	C	D	E	F
w_Md	A	b	c	d	e	F
w_Mp	a	(b)	(c)	d	e	F
w_F	(a)	(b)	(c)	(d)	(e)	f
w_MF	inapt	inapt	inapt	inapt	(e)	(f)

Where IGs are Annual Crops (AC), Perennial Crops (PC), Grassland(G), Agroforestry Systems (AFS). Forest Plantation (FP) and Natural Forestry (NF) and

QC are Good (capital letters), Moderate (lowercase letters), Restricted (lowercase letters in brackets) and Inapt.

Table 4.8 Framework for Quality Classes (QC) inside each Intensity Group (IG) and Technological Level (TL) in relation to Oxygen Deficiency Limiting Factors (LF)

LF	AC	CP	AFS	G	FP	NF
Technological Level: 1						
o_N	A	B	C	D	E	F
o_L	a	b	c	D	E	F
o_M	a	b	c	D	E	F
o_F	inapt	inapt	(c)	d	e	F
o_MF	inapt	inapt	inapt	(d)	(e)	f
Technological Level: 2						
o_N	A	B	C	D	E	F
o_L	a	b	c	E	E	F
o_M	(a)	b	c	E	E	F
o_F	(a)	(b)	(c)	d	e	F
o_MF	inapt	inapt	inapt	(d)	(e)	f
Technological Level: 3						
o_N	A	B	C	D	E	F
o_L	a	b	c	D	E	F
o_M	a	b	c	D	E	F
o_F	(a)	(b)	(c)	d	e	F
o_MF	(a)	(b)	(c)	d	e	f

Where IGs are Annual Crops (AC), Perennial Crops (PC), Grassland(G), Agroforestry Systems (AFS). Forest Plantation (FP) and Natural Forestry (NF) and QC are Good (capital letters), Moderate (lowercase letters), Restricted (lowercase letters in brackets) and Inapt.

Table 4.9 Framework for Quality Classes (QC) inside each Intensity Group (IG) and Technological Level (TL) in relation to Erosion Susceptibility Limiting Factors (LF)

LF	AC	CP	AFS	G	FP	NF
			Technological Level: 1			
e_N	A	B	C	D	E	F
e_L	a	B	C	D	E	F
e_M	(a)	b	c	d	E	F
e_F	(a)	(b)	c	d	e	F
e_MF	inapt	(b)	(c)	(d)	e	F
			Technological Level: 2			
e_N	A	B	C	D	E	F
e_L	A	B	C	D	E	F
e_M	a	b	c	D	E	F
e_F	(a)	b	c	d	E	F
e_MF	inapt	(b)	(c)	(d)	e	F
			Technological Level: 3			
e_N	A	B	C	D	E	F
e_L	A	B	C	D	E	F
e_M	a	b	b	D	E	F
e_F	(a)	b	b	c	E	F
e_MF	inapt	(b)	(b)	(c)	e	F

Where IGs are Annual Crops (AC), Perennial Crops (PC), Grassland(G), Agroforestry Systems (AFS). Forest Plantation (FP) and Natural Forestry (NF) and QC are Good (capital letters), Moderate (lowercase letters), Restricted (lowercase letters in brackets) and Inapt.

Table 4.10 Framework for Quality Classes (QC) inside each Intensity Group (IG) and Technological Level (TL) in relation to Impediments to Mechanization Limiting Factors (LF)

LF	AC	CP	AFS	G	FP	NF
			Technological Level: 1			
m_N	A	B	C	D	E	F
m_L	A	B	C	D	E	F
m_M	A	B	C	D	E	F
m_F	a	d	g	j	E	F
m_MF	(a)	(d)	(g)	(j)	m	F
			Technological Level: 2			
m_N	A	B	C	D	E	F
m_L	A	B	C	D	E	F
m_M	b	B	C	D	E	F
m_F	b	e	h	D	E	F
m_MF	inapt	(e)	(h)	k	n	q
			Technological Level: 3			
m_N	A	B	C	D	E	F
m_L	a	B	C	D	E	F
m_M	(a)	b	c	D	E	F
m_F	(a)	(b)	(c)	d	e	f
m_MF	inapt	inapt	inapt	(d)	(e)	(f)

Where IGs are Annual Crops (AC), Perennial Crops (PC), Grassland(G), Agroforestry Systems (AFS). Forest Plantation (FP) and Natural Forestry (NF) and QC are Good (capital letters), Moderate (lowercase letters), Restricted (lowercase letters in brackets) and Inapt.

Table 4.11 Land Use Recommended for Technological Level Number 1

LF	AC	CP	AFS	G	FP	NF
Fertility	insert here the QC defined for each IG in the 1TL by Table 4.6					
Water	insert here the QC defined for each IG in the 1TL by Table 4.7					
Oxygen	insert here the QC defined for each IG in the 1TL by Table 4.8					
Erosion	insert here the QC defined for each IG in the 1TL by Table 4.9					
Mechanization	insert here the QC defined for each IG in the 1TL by Table 4.10					
Land Use	For each IG column, copy the most restrictive QC from the lines above					

Table 4.12 Land Use Recommended for Technological Level Number 2

LF	AC	CP	AFS	G	FP	NF
Fertility	insert here the QC defined for each IG in the 2TL by Table 4.6					
Water	insert here the QC defined for each IG in the 2TL by Table 4.7					
Oxygen	insert here the QC defined for each IG in the 2TL by Table 4.8					
Erosion	insert here the QC defined for each IG in the 2TL by Table 4.9					
Mechanization	insert here the QC defined for each IG in the 2TL by Table 4.10					
Land Use	For each IG column, copy the most restrictive QC from the lines above					

Table 4.13 Land Use Recommended for Technological Level Number 3

LF	AC	CP	AFS	G	FP	NF
Fertility	insert here the QC defined for each IG in the 3TL by Table 4.6					
Water	insert here the QC defined for each IG in the 3TL by Table 4.7					
Oxygen	insert here the QC defined for each IG in the 3TL by Table 4.8					
Erosion	insert here the QC defined for each IG in the 3TL by Table 4.9					
Mechanization	insert here the QC defined for each IG in the 3TL by Table 4.10					
Land Use	For each IG column, copy the most restrictive QC from the lines above					

Table 4.14 Recommended Land Use

LF	AC	CP	AFS	G	FP	NF
1 TL	insert the last line of table 4.11 here					
2 TL	insert the last line of table 4.12 here					
3 TL	insert the last line of table 4.13 here					
Land Use	For each IG column, copy the least restrictive QC from the lines above, also indicating the corresponding technological level in front of each QC					

5 ADEQUACY OF RECOMMENDED LAND USE

"The Ogallala aquifer is one of the world's largest underground sources of freshwater. For decades, farmers and others have been slurping up groundwater far faster than nature can recharge it". (Malewitz e Writer 2013).
If we have a limited resource, we cannot think that its unbridled use will have no cost for this generation or future generations. We must do something to protect it.

The Adequacy of Recommended Land Use (ARL) consists of a comparison between the land use indicated by the Recommendation for Sustainable Land Use (ReSLU) and the most recently practiced land use.

This comparison is made using an ordinal scale ranging from -5 to + 5, where the median value (zero) represents a situation where the most recent land use matches the land use indicated by ReSLU.

Positive values indicate that the current land use intensity remains below levels that cause land degradation. This condition allows for soil and water conservation in addition to other benefits, such as the protection of fauna and flora at higher indicator levels (+ 4 or +5).

For example, if the recommended land use for a particular technological level is good for annual crops, and the current use is natural forest, the indicator value will be +5. This implies a higher degree of environmental conservation for the management unit, since lands that are highly resistant to degradation are kept in a state of preservation.

Negative values reveal that the recent land use intensity surpasses what the land can support. Under these land use conditions, the land suffers environmental degradation that manifests itself in soil and water losses, as well as the loss of biodiversity, when the indicator reaches lower values (-5 or -4). Negative values indicate the need to adjust land use toward more sustainable management practices. In addition, the smaller the value, the greater the urgency to change the land use.

For example, if the recommended land use is grasses and the most recent land use is annual crops, the degree of suitability will be -2. Therefore, results indicate the need to adopt a less intensive land use, such as pasture, agroforestry systems or plantation forests.

The ARL index is obtained using a weighted matrix (Table 5.1), where the current land use is compared with the recommended land use. The only restriction is that the perimeter of each land use class (recommended and current) fit within each other.

Table 5.1. ARL index represented by a weighted matrix between current land use and recommended land use.

		Recommended Land Use					
		Annual crops	Perennial crops	Grasses	Agroforestry systems	Plantation Forest	Natural Forests
Current Land Use	Annual crops	0	-1	-2	-3	-4	-5
	Perennial crops	1	0	-1	-2	-3	-4
	Grasses	2	1	0	-1	-2	-3
	Agroforestry systems	3	2	1	0	-1	-2
	Plantation Forest	4	3	2	1	0	-1
	Natural Forests	5	4	3	2	1	0

Since the ReSLU identifies specific factors that limit land use, we can easily identify the technologies or processes necessary to promote sustainable land management within each technological level.

For example, if the main limitation factor is impediment to mechanization under the 3TL category, a simple change of in land use from 3TL to 1TL can result in more sustainable land use, and could even qualify the land as an environmental services supplier.

Another example is when the main limitation factor for land use is soil acidity (fertility deficiency) within the 1TL category. Soil-liming could transform this land from 1TL to 3TL, resulting in a more sustainable land use and again qualifying the land as an environmental services provider.

The core concept of ARL is that by changing the land use type or the technologies used for management, more sustainable conditions can be achieved, and even qualify the land as a supplier of environmental services.

LITERATURE CITED

Alvarez V., V. H.; Novais, , R. F.; Dias, , L. E.; Oliveira, J. A. Determinação e uso do fósforo remanescente. Boletim Informativo da Sociedade Brasileira de Ciência do Solo, 25:27-32. 2000.

Andreoli, M., and V. Tellarini. Farm sustainability evaluation: methodology and practice. Agriculture, Ecosystems and Environment , 77:43–52. 2000.

Anjos, L. H. C.; Silva, L. M.; Wadt, P. G. S.; Lumbreras, J. F.; Pereira, M. G. Guia de Campo da IX Reunião Brasileira de Classificação e Correlação de Solos. 1. ed. Rio Branco: Embrapa / SBCS, 2013. 204p.

Arruda, F. B.; Zullo Junior, J.; Oliveira, J. B.; Parâmetros de solo para o cálculo da água disponível com base na textura do solo, Revista Brasileira de Ciência do Solo, 1:11-15, 1987.

Ávila, M. M. Avaliação ponderada de impacto ambiental em propriedades rurais do Estado do Acre, na Amazônia Brasileira. Rio Branco, AC: Dissertação (Ecologia e Manejo de Recursos Naturais). Universidade Federal do Acre, 2006. 49p.

Brady, M; Sahrbacher C.; Kellermann, K; and Happe, K. An agent-based approach to modeling impacts of agricultural policy on land use, biodiversity and ecosystem services. Landscape Ecology, 27: 1363–1381. 2012.

Bronner, E. Workers Claim Race Bias as Farms Rely on Immigrants. New York Times, 2013.

<http://www.nytimes.com/2013/05/07/us/suit-cites-race-bias-in-farms-use-of-immigrants.html> (accessed May 7, 2013).

Campêlo, G. Projetos extrativistas em florestas fracassam, aponta Banco Mundial. Ambiental Sustentável. <http://ambientalsustentavel.org/2013/projetos-extrativistas-em-florestas-fracassam-aponta-banco-mundial/> (accessed April, 15, 2013).

Claassen, R; Cattaneob A.; and Robert J. Cost-effective design of agri-environmental payment programs: U.S. experience in theory and practice. Ecological Economics, 65: 737 – 752. 2008.

Delarmelinda, E. A. ; Wadt, P. G. S. ; Anjos, L. H. C. ; Masutti, C. S. M. ; Silva, E. F. ; Silva, M. B. E. ; Coelho, R. M. ; Shimizu, S. H. ; Couto, W. H. . Avaliação da Aptidão Agrícola dos Solos do Acre por Diferentes Especialistas. Revista Brasileira de Ciência do Solo. 35:1841-1853. 2011.

Delarmelinda, E. A. Aplicação de sistemas de avaliação da aptidão agrícola em solos do Estado do Acre. Rio Branco, Acre: Dissertação (Produção Vegetal). Universidade Federal do Acre, 2011. 142p.

Dobbs, T. L.; Pretty, J. Case study of agri-environmental payments: The United Kingdom. Ecological Economics, 65: 765-775. 2008.

Ecological Society of America. Ecosystem Services: A Primer. Ecological Society of America. <http://www.actionbioscience.org/environment/esa.html> (accessed January, 13, 2013).

Empresa Brasileira de Pesquisa Agropecuária - EMBRAPA. Centro Nacional de Pesquisa de Solos. Manual de métodos de análise de solo. 2.ed. Rio de Janeiro, Embrapa Solos, 1997, 212p.

Empresa Brasileira de Pesquisa Agropecuária - EMBRAPA. Sistema Brasileiro de Classificação de Solos. 2.ed. Rio de Janeiro, 2006. 306p.

European Commission Directorate General for Agriculture and Rural Development. Agri-environment Measures. Overview on General Principles, Types of. Unit G-4 - Evaluation of Measures applied to Agriculture, Studies , 2005.

Greenhouse, S. Major Retailers Join Bangladesh Safety Plan. New York Times, 2013.

<http://www.nytimes.com/2013/05/14/business/global/hm-agrees-to-bangladesh-safety-plan.html> (accessed May, 29, 2013).

Haaren, C. von; Kempa, D; Vogel, K; and Rüter; S. Assessing biodiversity on the farm scale as basis for ecosystem service payments. Journal of Environmental Management, 113: 40-50, 2012.

Hinnant, L.; and Borenstein, S. Scientists find universe is 80 million years older. Time. march 21, 2013. http://science.time.com/2013/03/21/universe-ages-80m-years-big-bang-gets-clearer/ (accessed march 21, 2013).

Holand, R. A.; Eigenbrod, F.; Armsworth, P. R ; Anderson, B. J.; Thomas, C. D. and Gaston, K. J. "The influence of temporal variation on relationships between ecosystem services." Biodivers Conserv 20: 3285–3294. 2011.

Lepsch, I.F. Manual para levantamento utilitário do meio físico e classificação de terras no sistema de capacidade de uso. Campinas: SBCS, 1991. 175p.

Macaulay Land Use Research Institute. Exploring Scotland. Land Capability for Agriculture. <http://www.macaulay.ac.uk/explorescotland/lca.html> (Accessed March, 27, 2013).

Maciel, R. C. G.; Reydon, B. P.; Costa, J. A.; and Sales, G. O. O.. Pagando pelos Serviços Ambientais: Uma proposta para a Reserva Extrativista Chico Mendes. Acta Amazônica 3: 489 - 498. 2010.

Malewitz, J, In drought Ravaged Plains, efforts to save a vital aquifer. The Pew Charitable Trusts. 17, 2013. http://www.pewstates.org/projects/stateline/headlines/in-drought-ravaged-plains-efforts-to-save-a-vital-aquifer-85899460061 (accessed March 24, 2013).

Manitoba Agriculture, Food and Rural Initiatives. Soil Survey: The Soil Landscapes of Manitoba. <http://www.gov.mb.ca/agriculture/soilwater/soilsurvey/index.htm l> (accessed April, 16, 2013);

Mann, M. L.; Kaufmann, R. K; Bauer, D. M.; Sucharita, G.; James, B. G.; and Vera-Diaz, M. C.. Ecosystem Service Value and Agricultural

Conversion in the Amazon: Implications for Policy Intervention. Environ Resource Econ 53:279–295. 2012.

Pagiola, S.; Arcenas, A.; Platais, G.. Can Payments for Environmental Services Help Reduce Poverty? An Exploration of the Issues and the Evidence to Data from Latin America. World Development, 237-253, 2005.

Ramalho Filho, A. and Beek, K.J. Sistema de avaliação da aptidão agrícola das terras. 3.ed. Rio de Janeiro, EMBRAPA, 1995. 65p.

Riley, N. Florida´s Farmworkers in the twenty-first century. University Press of Florida. 208p 2002

Rodrigues, G. S., and Campanhola, C. Sistema integrado de avaliação de impacto ambiental aplicado a atividades do Novo Rural. Revista Pesquisa Agropecuária Brasileira 38: 445-451. 2003.

Smith, C. S. and McDonald, G. T. Assessing the sustainability of agriculture at the planning stage. Journal of Environmental Management, no. 52: 15–37. 1988.

Song, Y.; Lianyou, L.; Ping, Y. and Tong, C. A. O. A review of soil erodibility in water and wind erosion research. Journal of Geographical Sciences, 15: 167-176. 2005.

U.S. Department of Agriculture, Natural Resources Conservation Service. National soil survey handbook, title 430-VI. http://soils.usda.gov/technical/handbook/ (accessed March, 23, 2013).

Wang, G.; Innes, J. L.;Wu, S. W.; Krzyzanowski, J.; Yin, Y.. Dai, S.; Zang, X.; Liu, S. National park development in China: Conservation or commercialization? Ambio, 41: 247-261. 2012.

WEDU. "Adapted of the Martha Speaks TV Program." Florida West Coast Public Broadcasting Inc. (WEDU). All rights reserved , 2013.

Wikipedia. Kaesong Industrial Region. Wikipedia, 2013. <http://en.wikipedia.org/wiki/Kaesong_Industrial_Region> (accessed June 23, 2013),

ABOUT THE AUTHOR

Agricultural Engineer, Master of Soil Science and Ph.D. in Soil Science and Plant Nutrition. Brazilian Agricultural Research Corporation (Embrapa) researcher and professor of the graduate programs at the Federal University of Acre and at the Federal University of Amazonas, Brazil.

This book was written during the author's postdoctoral training at the University of Florida, with financial support from National Counsel of Technological and Scientific Development (CNPq) through Brazil's Science Without Borders Program.